Lecture Notes in Physics

220

Walter Dittrich
Martin Reuter

Effective Lagrangians in Quantum Electrodynamics

Springer-Verlag Berlin Heidelberg GmbH

Authors

Walter Dittrich
Martin Reuter
Institut für Theoretische Physik der Universität Tübingen
D-7400 Tübingen, F.R.G.

ISBN 978-3-540-15182-1 ISBN 978-3-540-39259-0 (eBook)
DOI 10.1007/978-3-540-39259-0

Library of Congress Cataloging in Publication Data. Dittrich, Walter. Effective Lagrangians in quantum electrodynamics. (Lecture notes in physics; 220) Bibliography: p. 1. Quantum electrodynamics. 2. Lagrangian functions. I. Reuter, Martin, 1958-. II. Title. III. Series. QC680.D53 1985 537.6 85-2527
ISBN 978-3-540-15182-1

Originally published by Springer-Verlag Berlin Heidelberg in 1985

2153/3140-543210

PREFACE

With these notes we would like to provide an introduction to the subject of effective Lagrangians for quantum electrodynamics (QED). Although this topic is interesting in its own right, it also provides us with an example of several important calculational techniques of QED which are usually not found in standard texts. Moreover, studying vacuum problems in QED, where matters are fairly well understood, can be a helpful preparation for similar considerations in quantum chromodynamics and other more complicated theories. In contrast to the latter, electrodynamics has the advantage that many calculations can still be done analytically.

To make our computations as transparent as possible, many of them are presented in great detail. In particular, this is true for the 2nd, 3rd and 4th sections. The reader who is mainly interested in general concepts could omit the rather technical derivations of these sections in a first reading.

We wish to thank Christel Kienle for her endless patience and skill in typing the various versions of the manuscript.

Tübingen, September 1984

W. Dittrich
M. Reuter

TABLE OF CONTENTS

(1) INTRODUCTION

The problem of the existence of a stable electron dates back to the very beginning of electrodynamics: if it is assumed to be an extended charge distribution, it is unstable due to the repulsive electrostatic forces, and if one assumes a point charge, one finds a divergent self energy.

Already at the beginning of this century, attempts were made to solve this problem by generalizing Maxwell's equations (Mie 1912, Born and Infeld 1934)*. Because these equations can be derived via a variational principle from a Lagrangian density L, it is natural to generalize the expression for L. In doing so, the following points must be taken into account:

(i) In order to generate Lorentz covariant equations, L must be a Lorentz scalar, i.e., it must be a function of invariant combinations of the field quantities.

(ii) L must be gauge invariant.

(iii) In the limit of small field strengths, L has to approach $L^{(o)} = \frac{1}{2}(\vec{E}^2-\vec{B}^2)$, which is the Lagrangian leading to Maxwell's equations.

The electromagnetic field has only two gauge invariant Lorentz scalars, viz.

$$\mathcal{F} = \frac{1}{4} F_{\mu\nu} F^{\mu\nu} = \frac{1}{2}(\vec{B}^2 - \vec{E}^2)$$

$$\mathcal{G}^2 = (\frac{1}{4} F_{\mu\nu} {}^*F^{\mu\nu})^2 = (\vec{E}\cdot\vec{B})^2$$

---

*The original papers of this section are cited in the list of references under ref. [1].

where ${}^{*}F^{\mu\nu} = \frac{1}{2}\,\varepsilon^{\mu\nu\rho\sigma}\,F_{\rho\sigma}$ is the dual field strength tensor. (note that $\vec{E}\cdot\vec{B}$ is a pseudoscalar, which changes its sign under a parity transformation). Thus, L can be a function of F and G^2 only. Born and Infeld used the following, quite arbitrary function of the invariants as their Lagrangian:

$$\mathcal{L} = E_0^2\left[1 - \left\{1 - \frac{E^2 - B^2}{E_0^2} - \frac{(\vec{E}\cdot\vec{B})^2}{E_0^4}\right\}^{1/2}\right]$$

Thereby, E_0 has the meaning of a maximum field strength. For fields much weaker than E_0, one recovers the ordinary Maxwell Lagrangian by expanding the square root and keeping only terms of second order. This function was chosen in analogy to the relativistic Lagrangian of a free particle, $L = mc^2[1-(1-v^2/c^2)^{1/2}]$, which reduces to the non-relativistic formula $L = \frac{1}{2}\,mv^2$ for $v \ll c$. Just as the limiting velocity $v = c$ is of no importance in classical mechanics, the maximum field strength E_0 is irrelevant for classical electrodynamics.

Indeed this non-linear theory could account for a stable electron with a finite size and finite self energy; however, other notorious difficulties within classical electrodynamics, such as the self-acceleration of charged particles, could not be resolved.

Whereas these first attempts to construct a non-linear electrodynamics were highly speculative, during the development of relativistic quantum mechanics in the thirties it became apparent that quantized matter fields can give rise to non-linear electromagnetic effects. This is easiest to understand in the context of quantum electrodynamics (QED), the relativistic quantum theory of the interaction between electrons, positrons and the electromagnetic field. Electromagnetic fields can be both macroscopic (external) fields and radiation fields (light quanta). It is necessary to formulate this theory as a quantum field theory. This means:

(i) All particles are described as excitation states of fields; thus, there is an electromagnetic field as well as a fermi field for the electrons and positrons in the theory.

(ii) The fields are not ordinary functions of space and time, but are non-commuting operators. This gives rise to uncertainty relations: If the field is precisely known in a point, its conjugate momentum is completely arbitrary. In general, the product of the uncertainties is given by Planck's constant.

Whether it is possible to use the classical concept of a particle or a field depends on the physical situation under consideration. For the moment, let us imagine a radiation field. This can be visualized as a system of an infinite number of coupled oscillators (one oscillator at each point of space) [21]. By introducing normal-coordinates, the system decouples and one gets an infinite set of free oscillators with the Hamiltonian

$$H = \frac{1}{2} \sum_i \left(p_i^2 + \omega_i^2 q_i^2 \right)$$

where p_i, q_i and ω_i denotes momentum, amplitude and frequency, respectively, of the i-th oscillator. Quantizing this system, one obtains for the possible energy eigenvalues

$$E = \sum_i \hbar \omega_i \left(N_i + \frac{1}{2} \right)$$

where N_i denotes the number of quanta in the mode i. Now let us look a little closer at the vacuum state. In a classical theory, there would not be much to say: the ground-state simply is the state with vanishing field strength all over the space. In quantum theory, matters are not so simple, because even if $N_i = 0$ for all modes i, we are still left with the zero-point fluctuations of all the harmonic oscillators, which contribute an energy $E_0 = \sum_i \frac{1}{2} \hbar\omega_i$. Because the number of modes is infinite, this zero-point energy diverges. Usually E_0 is eliminated by appropriately choosing the energy scale. However, there are observable consequences of this non-trivial structure of the (photon-) vacuum: consider, for example, the Casimir effect [20]. As was shown by Casimir, two uncharged, ideal conducting plates attract each other because of the fluctuations of the electromagnetic field. This point will be discussed in more detail in Appendix C.

Other observable consequences of the vacuum fluctuations of the photon field are the Lamb shift in atomic levels or the anomalous magnetic moment of the electron.

Thus we are forced to give up the concept of the vacuum being a particle and field free space. This is not only true for the radiation field, but also for the electron-positron, or, more generally, for every matter field. This means that the vacuum does not only contain local fluctuations of the electromagnetic field strengths (which can be interpreted as virtual photons), but also charge fluctuations due to the creation and the subsequent annihilation of electron-positron pairs. It is these charge fluctuations of the quantized electron-positron (Dirac) field with an external (i.e. unquantized) electromagnetic field which we now want to study in some detail. In the source-free regions of space this external field obeys in absence of matter fields the classical Maxwell equations $\partial_\mu F^{\mu\nu} = 0$, which, as stated above, can be derived from the variational principle

$$\frac{\delta W^{(0)}[A]}{\delta A^\mu(x)} \equiv \frac{\delta}{\delta A^\mu(x)} \int d^4x' \, \mathcal{L}^{(0)}(x') = 0 \quad ,$$

$$\mathcal{L}^{(0)} = -\frac{1}{4} F_{\mu\nu} F^{\mu\nu} \quad , \quad F_{\mu\nu} \equiv \partial_\mu A_\nu - \partial_\nu A_\mu \tag{1.1}$$

Our aim is to find an effective action $W_{eff}[A] = W^{(o)}[A] + W^{(1)}[A]$, where $W^{(1)}$ describes the non-linear effects induced by the quantized fermion fields. The new equations of motion are then given by

$$\frac{\delta W_{eff}[A]}{\delta A_\mu(x)} = 0 \tag{1.2}$$

To be precise, for the fermions we assume Dirac particles [37,39] described by the Lagrangian (for our conventions, see appendix A)

$$\mathcal{L}_\psi = \mathcal{L}_\psi^o + \mathcal{L}_\psi^W \tag{1.3}$$

$$\mathcal{L}_\psi^o = - \bar{\psi} \gamma^\mu \left(\frac{1}{i} \partial_\mu + m \right) \psi$$

$$\mathcal{L}_\psi^W = e \, \bar{\psi} \gamma^\mu \psi \, A_\mu \equiv j^\mu A_\mu$$

The current j^μ of the electrons and positrons can be obtained from the action $W_\psi = \int d^4x \, L_\psi$ via

$$\frac{\delta W_\psi[\psi, \bar{\psi}, A]}{\delta A_\mu(x)} = j^\mu(x) \tag{1.4}$$

It is this current to which the external field $A^\mu(x)$ couples; however, we are not actually interested in a theory which explicitly contains the Dirac field ψ. If we wanted that, we would stop after having derived (1.3). Instead, we are looking for an action functional $W^{(1)}[A]$ which simulates the presence of the electrons at a purely classical level. If one now defines $W^{(1)}$ so that it generates the vacuum expectation value of j^μ upon differentiating with respect to A_μ, i.e. by

$$\frac{\delta W^{(1)}[A]}{\delta A_\mu(x)} = \langle 0 | j^\mu(x) | 0 \rangle^A \tag{1.5}$$

the generalized Maxwell equations (1.2) read

$$\partial_\nu F^{\nu\mu}(x) = \langle 0 | j^\mu(x) | 0 \rangle^A \tag{1.6}$$

Because we have averaged over ψ, these are (in general highly non-linear) equations of motion for the A_μ field only. Because of $\partial_\mu \partial_\nu F^{\nu\mu} = 0$ ($F^{\mu\nu}$ is antisymmetric!), the vacuum current must be conserved for (1.6) to be consistent:

$$\partial_\mu \langle 0 | j^\mu(x) | 0 \rangle = 0 \tag{1.7}$$

This is indeed the case if $W^{(1)}$ is a gauge invariant functional of A_μ, i.e., if we have

$$W^{(1)}[A^\Lambda] = W^{(1)}[A] \tag{1.8}$$

where A^Λ is a gauge transform of A:

$$A_\mu^\Lambda = A_\mu - \partial_\mu \Lambda \tag{1.9}$$

Using the chain rule, (1.8) can be exploited as follows

$$\begin{aligned} 0 &= \frac{\delta}{\delta\Lambda(x)} W^{(1)}[A^\Lambda]\Big|_{\Lambda=0} \\ &= \int d^4y \, \frac{\delta W^{(1)}[A^\Lambda]}{\delta A_\mu^\Lambda(y)}\Big|_{\Lambda=0} \frac{\delta A_\mu^\Lambda(y)}{\delta\Lambda(x)} \\ &= -\int d^4y \, \langle 0| j^\mu(y) |0\rangle \, \partial_\mu^y \delta(y-x) \\ &= \partial_\mu \langle 0| j^\mu(x) |0\rangle \end{aligned} \tag{1.10}$$

In the third line, (1.5) was used. Now we have shown that a gauge invariant effective action functional leads to physically acceptable equations of motion and our remaining task is to solve eq. (1.5) together with the boundary condition $W^{(1)}\ [F_{\mu\nu} = 0] = 0$ for $W^{(1)}$.

To this end we first redefine the current appearing on the right-hand side of (1.6). As is well kown [37], when quantizing the electron field by imposing the anti-commutation relations

$$\{\psi_\alpha(\vec{x},t), \overline{\psi}_\beta(\vec{y},t)\} = \gamma^0_{\alpha\beta}\, \delta(\vec{x}-\vec{y}) \tag{1.11}$$

one obtains an infinite total charge for the vacuum state. Subtracting this infinite quantity from the charge operator corresponds to replacing $\bar{\psi}\gamma^\mu\psi$ by the current

$$j^\mu = e\,\bar{\psi}\gamma^\mu\psi - \frac{e}{2}\{\bar{\psi},\gamma^\mu\psi\}$$
$$= \frac{e}{2}[\bar{\psi},\gamma^\mu\psi] \qquad (1.12)$$

which is gauge invariant and fulfills

$$\langle 0|j^\mu(x)|0\rangle^{A_{\mu\nu}=0} = 0 \qquad (1.13)$$

The vacuum expectation value now reads

$$\langle 0|j^\mu(x)|0\rangle^A = \frac{e}{2}\lim_{x'\to x}\gamma^\mu_{\alpha\beta}\langle 0|\bar{\psi}_\alpha(x)\psi_\beta(x') - \psi_\beta(x')\bar{\psi}_\alpha(x)|0\rangle$$
$$= -e\lim_{\substack{x'\to x\\ S}}\gamma^\mu_{\alpha\beta}\langle 0|T\,\psi_\beta(x')\bar{\psi}_\alpha(x)|0\rangle$$
$$(1.14)$$

Hereby T denotes the time ordering operator

$$T\,A(x)B(x') = \Theta(x^0-x'^0)A(x)B(x') - \Theta(x'^0-x^0)B(x')A(x) \qquad (1.15)$$

and the coincidence limit has to be performed symmetrically with respect to the time coordinate:

$$\lim_{\substack{x'\to x\\ S}} \equiv \frac{1}{2}\left(\lim_{\substack{x'\to x\\ x^{0\prime}>x^0}} + \lim_{\substack{x'\to x\\ x^{0\prime}<x^0}}\right) \qquad (1.16)$$

Now it is convenient to introduce the fermion Green's function (propagator, two-point function) defined by

$$G_{\beta\alpha}(x,x') = i\langle 0|T\,\psi_\beta(x)\bar{\psi}_\alpha(x')|0\rangle \qquad (1.17)$$

which is the resolvent of the Dirac operator [37]:

$$[(\gamma^\mu \Pi_\mu)_{\alpha\beta} + m\,\delta_{\alpha\beta}]\, G_{\beta\gamma}(x,x') = \delta_{\alpha\gamma}\,\delta(x-x') \tag{1.18}$$

with

$$\Pi_\mu = \frac{1}{i}\partial_\mu - e A_\mu \tag{1.19}$$

(Recall that the Dirac equation resulting from (1.3) reads $[\gamma^\mu \Pi_\mu + m]\psi = 0$). Hence the current expectation value is given by

$$\begin{aligned} \langle 0 | j^\mu(x) | 0 \rangle^A &= ie \lim_{x' \xrightarrow{s} x} \gamma^\mu_{\alpha\beta}\, G_{\beta\alpha}(x,x') \\ &\equiv ie\, \mathrm{tr}[\gamma^\mu G(x,x)] \end{aligned} \tag{1.20}$$

where the symmetric limit is understood in the sequel. The defining equations for $W^{(1)}$ now are

$$\begin{aligned} \frac{\delta W^{(1)}[A]}{\delta A_\mu(x)} &= ie\, \mathrm{tr}[\gamma^\mu G(x,x)] \\ W^{(1)}[F_{\mu\nu} = 0] &= 0 \end{aligned} \tag{1.21}$$

As is demonstrated in appendix D, this problem is solved by

$$\begin{aligned} W^{(1)}[A] &= i\, \mathrm{Tr}\, \ln (1 - e\,\gamma A\, G_+)^{-1} \\ &= i\, \mathrm{Tr}\, \ln \left(G_+[A] / G_+[0] \right) \end{aligned} \tag{1.22}$$

with $G_+[A]$ the propagator in an external field

$$G_+[A] = \frac{1}{\gamma\Pi + m - i\varepsilon} \tag{1.23}$$

which is connected with $G_+ \equiv G_+[F_{\mu\nu} = 0]$ by

$$G_+[A] = G_+(1 - e\,\gamma A\, G_+)^{-1} \tag{1.24}$$

The symbol Tr denotes $\int d^4x$ tr, i.e. the trace in both spinor and configuration space. In the language of Feynman diagrams, (1.21) and (1.22), respectively, are represented by a single electron loop in an external field (a "short-cut" propagator G[A]):

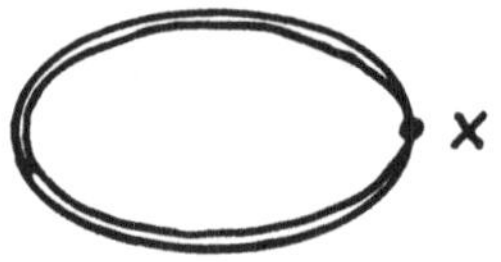

(1.25)

(The double line denotes the presence of an external field). The evaluation of (1.22) for a given potential $A_\mu(x)$ will, in general, be an extremely complicated task. Simple solutions are known only for a very limited class of fields (constant fields, laser fields, weak fields, slowly varying fields, etc.).

The first people who discussed effective actions like (1.22) were Heisenberg and Euler [1], as well as Weisskopf [1], in 1936. Then, in 1951, Schwinger [3] published a classical paper in which he evaluated $W^{(1)}[A]$ for several types of fields. He used the so-called proper-time method which reduces the calculation of (1.22) to a one-dimensional problem of ordinary particle quantum mechanics. In chapter (5), we will use a similar method to calculate $\mathcal{L}^{(1)}$ (frequently called Heisenberg-Euler Lagrangian) defined by

$$W^{(1)} = \int d^4x \, \mathcal{L}^{(1)} \tag{1.26}$$

for a constant magnetic field.

Now, looking back to the early works of Mie and Born and Infeld, we see that today the motivations for studying non-linear generalizations of Maxwell's equations for the vacuum are quite different. The main objective when studying effective actions is to learn something about the structure of the vacuum which, in this approach, is probed by an external electromagnetic field. However, from the modern point of view, the problem of a stable electron is not an issue which can be discussed in terms of pure electrodynamics; instead, it should be solved within an up to now unknown fundamental theory of matter and its interactions. The problem of the diverging self-energy, for example, is not really solved within the present theory but is hidden behind sophisticated renormalization schemes.

At this point it might be interesting to look at a system closely related to the quantized fermions in an external, i.e., classical, electromagnetic field, viz. a quantized matter field in presence of a classical background gravitational field. For simplicity's sake, we consider a free scalar field $\phi(x)$ with the classical action [48]

$$W_\phi = \frac{1}{2}\int d^4x \sqrt{-g(x)}\left\{-g^{\mu\nu}(x)\,\partial_\mu\phi\,\partial_\nu\phi - m^2\phi^2\right\} \qquad (1.27)$$

where the metric tensor field is a prescribed function of x. (The most general Lagrangian for ϕ would also contain a term $\sim R\phi^2$ with R being the scalar curvature). In a semiclassical theory where $g^{\mu\nu}(x)$ is treated classically whereas $\phi(x)$ is treated quantum mechanically, the vacuum expectation value of the matter field

$$T_{\mu\nu} = \partial_\mu\phi\,\partial_\nu\phi - \frac{1}{2}g_{\mu\nu}\,g^{\sigma\varsigma}\,\partial_\sigma\phi\,\partial_\varsigma\phi \qquad (1.28)$$

acts as a source on the right-hand side of Einstein's equations:

$$R_{\mu\nu} - \frac{1}{2} g_{\mu\nu} R + \Lambda g_{\mu\nu} = - 8\pi G \langle 0 | T_{\mu\nu} | 0 \rangle^{g} \tag{1.29}$$

Obviously, this is the analogue of (1.6). Eq. (1.29) with the right-hand side set equal to zero is obtained as the variation of the usual Einstein-Hilbert action

$$W^{(0)}[g] = \frac{1}{16\pi G} \int d^4x \sqrt{-g}\,(R - 2\Lambda) \tag{1.30}$$

as

$$\frac{2}{\sqrt{-g}} \frac{\delta W^{(0)}[g]}{\delta g^{\mu\nu}(x)} = 0 \tag{1.31}$$

If we now define the effective action $W^{(1)}$ by

$$\frac{2}{\sqrt{-g}} \frac{\delta W^{(1)}[g]}{\delta g^{\mu\nu}(x)} = \langle 0 | T_{\mu\nu}(x) | 0 \rangle \tag{1.32}$$

our generlized equations (1.29) are given by

$$\frac{2}{\sqrt{-g}} \frac{\delta}{\delta g^{\mu\nu}(x)} \left(W^{(0)} + W^{(1)} \right) = 0 \tag{1.33}$$

which is clearly the analogue of (1.2). However, in contrast to Maxwell's equations, Einstein's equations are highly non-linear already at the purely classical level. But the strategy is the same in both cases: because one does not want to treat the microscopic degrees of freedom (the quantum fields $\psi(x)$ or $\phi(x)$) explicitly, one derives effective equations of motion simulating their presence for the macroscopic, classical field $A_\mu(x)$ or $g_{\mu\nu}(x)$.

For the equations (1.29) to be consistent, $\langle 0|T_{\mu\nu}|0\rangle$ must be covariantly conserved, because the left-hand side of (1.29) is.

By a reasoning analogous to (1.10) one can show [49], that $\langle 0|T_{\mu\nu}|0\rangle$ is indeed covariantly conserved if $W^{(1)}$ is invariant under general coordinate transformations, just as it was necessary for $W^{(1)}$ of electrodynamics to be gauge invariant for the induced vacuum current to be conserved. This is one example of the correspondence between gauge invariance in electrodynamics and general covariance in gravitation theory. Finally we mention that also in this case, $W^{(1)}$ can be expressed via the matter field propagator:

$$W^{(1)}[g] = -\frac{1}{2}\, i\, \mathrm{Tr}\, \ln\left(G_+[g] / G_+ \right) . \tag{1.34}$$

Now $G_+[g]$ denotes the propagator in presence of a gravitational field described by $g_{\mu\nu}(x)$, and G_+ is the corresponding flat space-time propagator. (The factor of -1/2 which is not present in (1.22) is due to the fact that ϕ is an uncharged scalar field, whereas ψ was a Dirac field). For a comprehensive introduction to these questions, see [48].

After this digression, let us return to our original problem of quantized fermions in a classical background electromagnetic field, which we now reconsider from the path integral point of view (for an introduction, see [24,25]). For the moment, let us consider an arbitrary field theory with fields $\{\phi\}$ and Lagrangian $L(\{\phi\})$. Transition amplitudes then can be expressed as functional integrals of the general form

$$Z = \int [d\{\phi\}] \exp\left(i S[\{\phi\}] \right) \tag{1.35}$$

with the action $S[\{\phi\}] = \int d^4x\, L(\{\phi\})$. Now assume that the set

$\{\phi\}$ can be divided in two subsets $\{\phi^L\}$ and $\{\phi^H\}$, where $\{\phi^L\}$ are "light" field components whose dynamics we directly observe (the photon field, or the classical $A_\mu(x)$, in our case), while $\{\phi^H\}$ are "heavy" fields (the electron field in our case) which influence the dynamics of the light fields but are not directly observable [50]. Since the $\{\phi^H\}$ are hidden from view, it is convenient to write (1.35) in the form

$$\begin{aligned} Z &= \int [d\{\phi^L\} d\{\phi^H\}] \exp\left(i S[\{\phi^L\}, \{\phi^H\}]\right) \\ &= \int [d\{\phi^L\}] \exp\left(i W_{eff} [\{\phi^L\}]\right) \end{aligned} \tag{1.36}$$

where the effective action W_{eff} for the light fields is defined by

$$\exp\left(i W_{eff}[\{\phi^L\}]\right) = \int [d\{\phi^H\}] \exp\left(i S[\{\phi^L\}, \{\phi^H\}]\right) \tag{1.37}$$

Clearly, the effective action, if exactly known, gives a complete description of the dynamics of $\{\phi^L\}$ without any reference to the heavy fields.

Now let us consider several examples of such integrations over the heavy field components [50]. First of all, for the fermions in an external field we have

$$\begin{aligned} W_{eff} &= W^{(0)} + W^{(1)} \\ W^{(0)} &\equiv \int d^4x \left\{ -\frac{1}{4} F_{\mu\nu} F^{\mu\nu} \right\} \end{aligned} \tag{1.38}$$

and $W^{(1)}$ is given by (note that $\exp(iW^{(o)})$ cancels on both sides of (1.37))

$$\exp\left(iW^{(1)}[A]\right) =$$

$$\int [d\psi d\bar{\psi}] \exp\left(-i\int d^4x\, \bar{\psi}\left[\gamma^\mu\left(\tfrac{1}{i}\partial_\mu - eA_\mu\right) + m\right]\psi\right) \qquad (1.39)$$

(Recall (1.3)). According to the general rules for the path integral quantization of Fermi fields [12], ψ and $\bar{\psi}$ are anti-commuting classical fields forming a Grassmann algebra; hence a Gauss-type integral like (1.39) can be evaluated to be [12]

$$\begin{aligned} \exp\left(iW^{(1)}[A]\right) &= \det\left[\gamma^\mu\left(\tfrac{1}{i}\partial_\mu - eA_\mu\right) + m\right] \\ &= \det\left(G[A]^{-1}\right) \end{aligned} \qquad (1.40)$$

This gives

$$\begin{aligned} W^{(1)}[A] &= -i \ln \det\left(G[A]^{-1}\right) \\ &= +i \ln \det G[A] \\ &= +i \operatorname{Tr} \ln G[A] \end{aligned} \qquad (1.41)$$

where we used the (formal) identity $\det(\exp G) = \exp(\operatorname{Tr} G)$. Because action functionals are defined only up to a constant, we may exploit this freedom to replace (1.41) by

$$\begin{aligned} W^{(1)}[A] &= i \operatorname{Tr} \ln G[A] - i \operatorname{Tr} \ln G[0] \\ &= i \operatorname{Tr} \ln\left(G[A]/G[0]\right) \end{aligned} \qquad (1.42)$$

which is identical to the previously derived result (1.22) and thus vanishes for $F_{\mu\nu} = 0$. Obviously, the notion of integrating out unobserved degrees of freedom together with the rules for the integration over Grassmann fields leads us back to the results already derived in a more pedestrian manner.

As another example of an effective Lagrangian, we mention the four-fermion interaction of the type $L = \frac{G_F}{\sqrt{2}} J^\mu J_\mu$ in the Fermi theory of weak interactions. The current J_μ consists of a leptonic part ℓ_μ and a hadronic part h_μ. A typical contribution to ℓ_μ is, for instance

$$\bar{e}\gamma_\mu(1-\gamma_5)\nu_e$$

describing the destruction of a neutrino and creation of an electron. The terms appearing in L have all the graphical representation

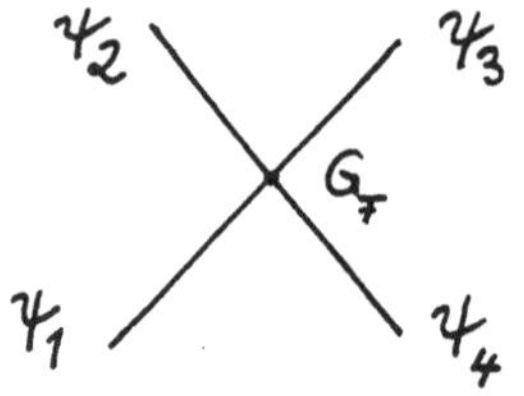

where the ψ_i's are arbitrary fermions ($e, \nu_e, \mu, \nu_\mu, \ldots$, quarks). Due to the fact that G_F has dimension $(\text{mass})^{-2}$, this field theory is non-renormalizable; nevertheless, it describes to a very good approximation weak interaction phenomena at low energies. As is generally believed, the "fundamental" theory of electro-weak interactions is the renormalizable Glashow-Weinberg-Salam gauge theory [51] in which, in addition to the fermions, the fundamental Lagrangian also contains gauge and Higgs bosons. The 4-fermion vertex is now replaced by the exchange of a heavy gauge boson $W^\pm$ or Z:

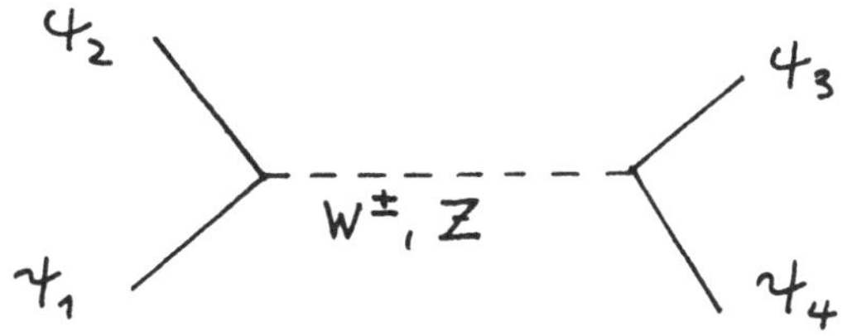

Because of the large mass of the gauge bosons, the forces mediated by them are very short-ranged; thus in the low energy limit (roughly $E < 80$ GeV), Fermi's point interaction is recovered. It is in this sense that the non-renormalizable $J_\mu J^\mu$-interaction can be regarded as an effective long wavelength or low energy effective Lagrangian of the renormalizable Glashow-Weinberg-Salam model. In a symbolic path-integral notation, this reads [50]:

$$\exp\left(i W_{eff}[\text{fermions}]\right) = \int [d\{\text{bosons}\}] \exp\left(i S_{\text{Weinberg-Salam}}[\{\text{fermions}\},\{\text{bosons}\}]\right) \tag{1.43}$$

Another possible application of the effective action concept is Adler's induced gravity approach to quantum gravity [50]. As has been long known, a quantum field theory of gravitation based upon the Einstein-Hilbert Lagrangian (1.30) is non-renormalizable because Newton's constant G has dimension $(\text{mass})^{-2}$. Now it is tempting to assume that there is some fundamental, renormalizable theory of gravitation which, upon integrating out unobserved matter fields, yields as an effective low energy (or long-wavelength) theory the Einstein-Hilbert action. At present, however, this approach is far from having been completely worked out; for a further discussion, the reader is invited to read the review article of Adler [50].

In the preceeding discussion we developed the intuitive notion of effective Lagrangians as describing the dynamics of "light" fields in interaction with "heavy" fields hidden from direct observation. But there is still another way to look at functionals like $W^{(1)}[A]$. As already explained, the fermion vacuum is characterized by a continuous creation and subsequent annihilation of (virtual) electron-positron pairs. Owing to the energy-time uncertainty principle $\Delta E \cdot \Delta t \gtrsim \hbar$, the maximum life-time of such a pair is about $\hbar/2mc^2$, where m is the electron's mass. If we apply a sufficiently strong external electric field to the vacuum, it is possible for this field to separate the electron from the positron so that no recombination takes place. In energetical terms this means that each of the particles must aquire an energy of at least mc^2 during its life-time $\hbar/2mc^2$. Then the virtual electron (positron) is converted into a real electron (positron). Of course, this is not a "creatio ex nihilo" because the energy corresponding to the rest-mass of the created particles is extracted from the external field. As we shall see in the later chapters, for the pair production rate to be significant, electric field strengths of about 10^{16} V/m are necessary; this tremendous number explains why one usually assumes the vacuum to be an insulator. In fact, at field strengths large enough, the vacuum becomes a conducting medium!

In other words, in presence of an external field the vacuum state $|0\rangle$ which contains no <u>real</u> particles can become unstable, i.e., it is energetically preferable that $|0\rangle$ decays into states containing real particles. If we prepare our system to be in the state $|0\rangle$ in the remote past $(t \to -\infty)$, then the probability amplitude $\langle 0, t = +\infty | 0, t = -\infty\rangle^A \equiv \langle 0_+ | 0_-\rangle^A$ for the system to remain in the ground-state $|0\rangle$ must not equal unity. Now, of course, the question arises: which is the functional dependence of the vacuum persistence amplitude $\langle 0_+ | 0_-\rangle^A$ on the external field described by the vector potential $A_\mu(x)$? One way to answer this question is to refer to standard texts on path-integral methods in field theory [25,26,12,51] where it is shown that this amplitude is given by exactly the path-integral (1.39), i.e., it is expressed by the effective action as

$$\langle 0_+ | 0_-\rangle^A = \exp\left(i W^{(1)}[A]\right) . \tag{1.44}$$

Thus, knowing $W^{(1)}$, we can calculate the probability of pairs being created as $1 - |\langle 0_+ | 0_-\rangle^A|^2$. This shows that only the imaginary part of $W^{(1)}$ is needed for this purpose.

The reader unfamiliar with path-integrals can derive (1.44) using standard perturbation theory [37,39]. In the interaction picture the relevant S-matrix element is given by

$$\langle 0_+ | 0_-\rangle^A = \langle 0 | S | 0\rangle = \langle 0 | T \exp\left(-i \int dt\, H_I\right) | 0\rangle \tag{1.45}$$

with the interaction Hamiltonian

$$H_I = \int d^3x \, \{ - j^\mu(x) A_\mu(x) \} \tag{1.46}$$

Expanding the exponential leads us to the perturbation series

$$\langle 0_+|0_-\rangle^A = \langle 0|T\exp\left(i\int d^4x\; j^\mu(x)\, A_\mu(x)\right)|0\rangle$$

$$= \sum_N \frac{i^N}{N!}\int d^4x_1 \ldots d^4x_N \;\; A_{\mu_1}(x_1) \ldots\ldots A_{\mu_N}(x_N) \tag{1.47}$$

$$\cdot \langle 0|T\, j^{\mu_1}(x_1) \ldots\ldots j^{\mu_N}(x_N)|0\rangle$$

(Recall that each $j^\mu(x)$ contains a factor of e). Applying Wick's theorem to the right-hand side of (1.47), we see that we have to sum up an infinite sequence of terms represented by diagrams like

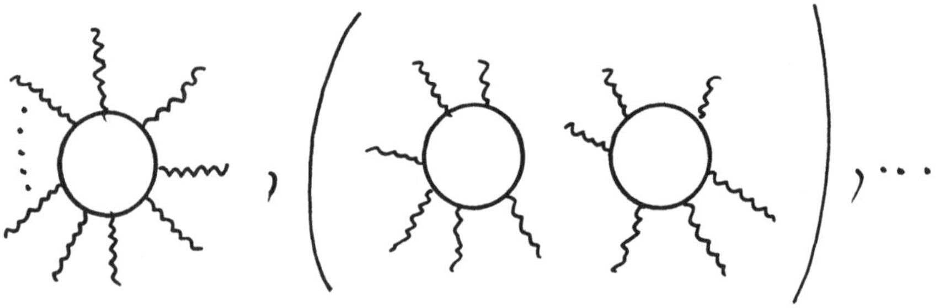

where the wavy lines denote interactions with the external field (no photons). This summation can be done explicitly [53] and the result is again (1.44). (Note that (1.47) also contains disconnected pieces like the second diagram above; owing to the "connectedness lemma" [41,25], these are not present in $W = -i \ln \langle 0_+|0_-\rangle^A$. The summation of the connected parts is carried out in appendix G).

Up to now, we always assumed that it is sensible to treat the electromagnetic field classically and only the fermion field quantum mechanically. Now let us ask how matters change if we also take the vacuum fluctuations of the photon field into account. This means that the total vector potential A_μ^{tot} now consists of two parts: $A_\mu^{tot}(x) = A_\mu(x) + a_\mu(x)$. As above, A_μ is a prescribed, classical background field; $a_\mu(x)$ denotes the fluctuations of the quantized photon field. If we calculate the effective action, or equivalently, the vacuum persistence amplitude in presence of these fluctuations, we have not only to integrate over the Dirac field but also over $a_\mu(x)$. Hence, (1.39) is replaced by

$$\langle 0_+|0_-\rangle^A = \int [d\psi d\bar\psi da] \exp\left(i\int d^4x \left[\mathcal{L}_a - \bar\psi\left(\frac{1}{i}\partial\!\!\!/ - eA\!\!\!/ - ea\!\!\!/ + m\right)\psi\right]\right) \tag{1.48}$$

with the photon kinetic term

$$\mathcal{L}_a = -\frac{1}{4} F_{\mu\nu} F^{\mu\nu} + \mathcal{L}_{gf} \quad , \quad F_{\mu\nu} \equiv \partial_\mu a_\nu - \partial_\nu a_\mu \tag{1.49}$$

where L_{gf} is a gauge-fixing term [25,12]. The integral (1.48) can be further evaluated using the following trick: one adds a term $j^\mu a_\mu$ to the Lagrangian and represents the a_μ field in the interaction term $\bar\psi\, a\!\!\!/\, \psi$ as $\frac{1}{i}\frac{\delta}{\delta j^\mu}$. Of course, at the end, one has to set j=0. This leads us to

$$\langle 0_+|0_-\rangle^A = \int [da] \int [d\psi d\bar\psi] \exp\left(-i\int d^4x\, \bar\psi\left(\frac{1}{i}\partial\!\!\!/ - eA\!\!\!/ - e\gamma^\mu \frac{1}{i}\frac{\delta}{\delta j^\mu(x)} + m\right)\psi\right) \cdot \exp\left(i\int d^4x \{\mathcal{L}_a + j^\mu a_\mu\}\right)\Bigg|_{j=0} \tag{1.50}$$

$$= \int [da] \det\left(\frac{1}{i}\partial\!\!\!/ - eA\!\!\!/ - e\gamma^\mu \frac{1}{i}\frac{\delta}{\delta j^\mu} + m\right) \cdot \exp\left(i\int d^4x \{\mathcal{L}_a + j^\mu a_\mu\}\right)\Bigg|_{j=0}$$

where we could perform the integral over ψ and $\bar{\psi}$ just as for $a_\mu = 0$. Now we may move the determinant in front of the remaining integral; this is easy to perform because $\int d^4x(L_a + j^\mu a_\mu)$ has the structure $\int d^4x(\frac{1}{2} a_\mu (D_+^{-1})^{\mu\nu} a_\nu + j^\mu a_\mu)$ with $D_+^{\mu\nu}$ being the photon propagator in the gauge specified by L_{gf}. Then, the a_μ-integral is of the Gauss-type and is readily evaluated by completing the square [12] . So we end up with

$$\langle 0_+|0_-\rangle^A = \det\left(\frac{1}{i}\not\partial - e\not A - e\gamma^\mu \frac{1}{i}\frac{\delta}{\delta j^\mu} + m\right) \cdot \exp\left(\frac{i}{2}\, jD_+j\right)\Big|_{j=0} \tag{1.51}$$

where we use a compact matrix notation:

$$jD_+j \equiv \int d^4x\, d^4y\; j_\mu(x)\, D_+^{\mu\nu}(x-y)\, j_\nu(y) \tag{1.52}$$

In writing down (1.51), we ignored a multiplicative constant; this corresponds simply to shifting W[A] by a meaningless (additive) constant. At this point, it is convenient to use the following identity [41] which is valid for any sufficiently differentiable functional F:

$$F\left[\frac{1}{i}\frac{\delta}{\delta j}\right] e^{\frac{i}{2} j\Delta j} = e^{\frac{i}{2} j\Delta j}\, e^{-\frac{i}{2}\frac{\delta}{\delta J}\Delta\frac{\delta}{\delta J}}\, F[J]\Big|_{J=\Delta j} \tag{1.53}$$

Again, $J = \Delta j$ means $J(x) = \int d^4y\, \Delta(x,y) j(y)$, etc..Thus, (1.51) becomes

$$\langle 0_+|0_-\rangle^A = \exp\left(\frac{i}{2}\, jD_+j\right) \exp\left(-\frac{i}{2}\frac{\delta}{\delta J} D_+ \frac{\delta}{\delta J}\right) \cdot \det\left(\frac{1}{i}\not\partial - e\not A - e\not J + m\right)\Big|_{j=0} \tag{1.54}$$

with $J = D_+ j$. Making use of (1.46), we obtain

$$\langle 0_+|0_-\rangle^A = \exp\left(-\frac{i}{2}\frac{\delta}{\delta J} D_+ \frac{\delta}{\delta J}\right) \exp\left(iW^{(1)}[A+J]\right)\Big|_{J=0} \tag{1.55}$$

This compact formula summarizes the effect of an arbitrary number of virtual photons (i.e., to all orders in perturbation theory) on the vacuum amplitude.

Clearly, it would be a formidable task to exactly evaluate (1.55) for an arbitrary $A_\mu(x)$; even for the restricted class of constant fields this is impossible, and one must restrict oneself to some approximation. In section (7), we will calculate the lowest-order corrections of $W^{(1)}$ for a pure magnetic field; for this case, it turns out that one has to compute only one further diagram beyond the one-loop graph (1.25) of $W^{(1)}$, namely, the two-loop diagram

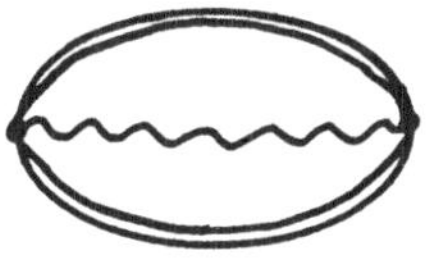

(1.56)

This gives rise to the so-called two-loop effective action $W^{(2)}$ or effective Lagrangian $L^{(2)}$. Such corrections were first studied by Ritus [4] and Dittrich [5], who obtain a rather complicated representation for $L^{(2)}(B)$. In chapter (7), we will derive a simpler form of $L^{(2)}(B)$. This calculation is based upon the observation that (1.56) can be written as the "convolution" of a photon with the polarization tensor of order α in the external field; symbolically:

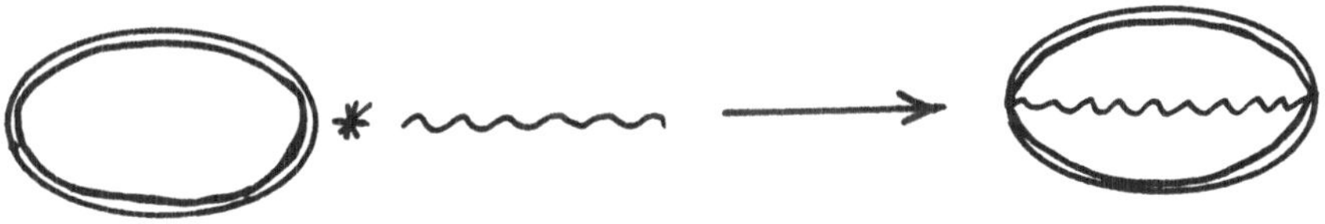

(1.57)

For the practical calculation, a momentum representation of the polarization tensor introduced by Tsai [6] proves very useful since it leads to a much more compact expression for the Lagrangian than the configuration space formulation in [4] and [5]. As we will see, the renormalization of $L^{(2)}$ also requires knowledge of the electron mass operator (to order α) for the external field; hereby we will again use a momentum representation given by Tsai [7].

However, before we attack the problem of calculating the various ingredients finally leading to $L^{(1)}$ and $L^{(2)}$, let us have a brief look at the various physical effects derivable from these Lagrangians. Because in this work we are mainly concerned with the techniques of calculating $L^{(1)}$ and $L^{(2)}$, the interested reader is referred to the review of Mitter [52], in which references to the original papers are also contained.
As was already mentioned, knowing the imaginary part of $W^{(1)}$, we can derive the rate of pair production in an external field. However, there are also several effects associated with the real part of $W^{(1)}$. These are similar to those appearing in dielectrics, i.e., media with dielectric constant $\varepsilon \neq 1$. A field dependent dielectric tensor ε_{Kl} $(\vec{E},\vec{B})$ can be assigned to the vacuum via the usual definition

$$D_k \equiv \frac{\partial \mathcal{L}_{eff}}{\partial E_k} = \varepsilon_{kl} E_l \qquad (1.58)$$

where $L_{eff} = L^{(o)} + L^{(1)} + L^{(2)}$ + higher corrections. Next, let us compile some of the effects arising from such non-linear dynamcis.

(i) Delbrück scattering

This means the scattering of a photon by a strong, slowly varying, external field, the Coulomb field of a nucleus, for instance. For nuclei with large charge Ze, the cross-section is of order mb, which is smaller than the corresponding Thomson cross section. In a Feynman diagram, such a process is represented as

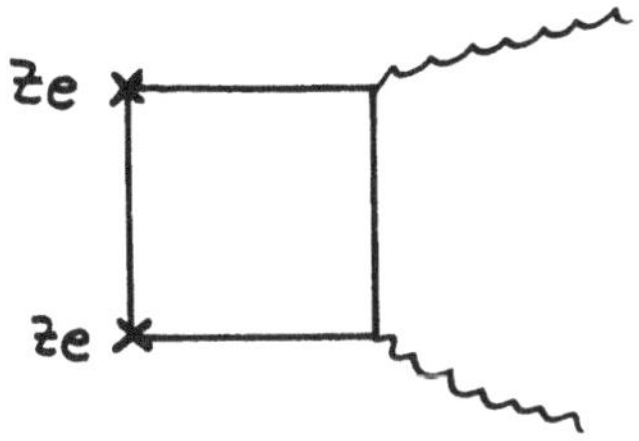

(ii) Double refraction

Due to the tensorial nature of ε (and analogously of the magnetic susceptibility μ), a strong magnetic or electric field can cause double refraction. If a light beam propagates perpendicular to the field direction, the phase velocity for the components polarized perpendicular to the field is different from that for the components polarized parallel to the field. This leads to two distinct indices of refraction.

(iii) photon splitting

This happens in presence of an external field and one or more photons: two photons superimpose to give one of a correspondingly higher energy or one photon splitts into two of lower energy. A typical diagram of such a process is the hexagon [2]:

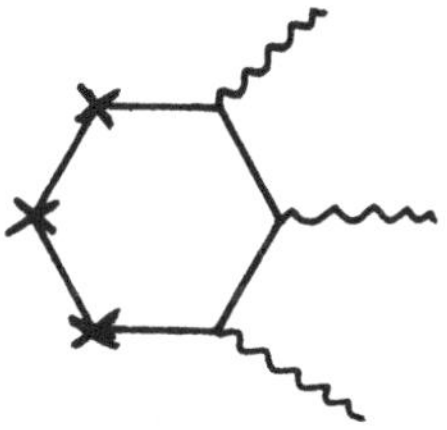

(1.59)

(The wavy lines are photons, the crosses external interactions). In absence of the external field, such processes are forbidden, due to Furry's theorem [53].

(iv) scattering of light by light

As was first shown by Heisenberg and Euler [1], processes like

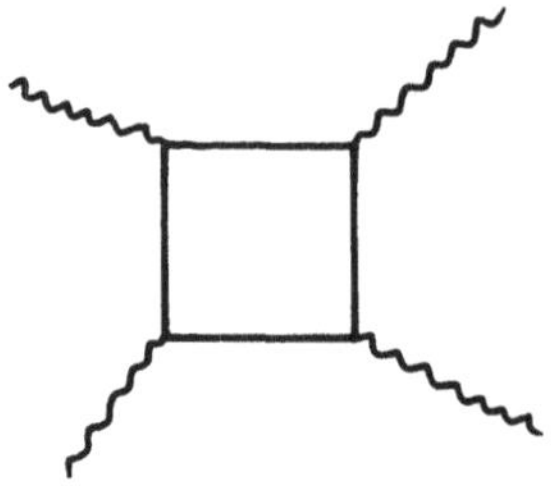

(1.60)

can lead to a photon-photon interaction [29]. This is a typical non-linear effect because in Maxwell's electrodynamics, there is no self-coupling of the fields. (This is reminiscent of the gluon-gluon coupling in non-Abelian gauge theories; in the latter, however, such couplings are already present at the tree level). The cross-section for (1.60) is of order μb.

Except for the Delbrück scattering, all the above effects are much to tiny to be measured experimentally even if the external field is the Coulomb field of a heavy nucleus. Inspite of their observation being problematic under usual conditions, these effects are interesting in their own right as an example of induced non-linear interactions; moreover, in the light of various recent attempts to formulate a theory of astrophysical objects surrounded by intense magnetic fields, it also becomes important to know the properties of the QED vacuum under such conditions.

With these remarks we end our introductory considerations on effective Lagrangians in general and the vacuum structure of QED in particular. The further sections of this study are organized as follows: First, in section (2), we derive in a self-contained manner an integral representation of the electron propagator in an external field, whereby we shall limit ourselves, as in all calculations which follow, to the case of a constant magnetic field. Building upon this, we will then construct the mass operator and the polarization tensor to oder e^2 in sections (3) and (4), thereby also deriving the conventional spectral representations as an additional illustration.

Then, in section (5), we derive an integral representation of the one-loop Lagrangian $L^{(1)}$, which is also needed to carry out the renormalization of $L^{(2)}$. This integral was formally solved by Dittrich [8] and numerically evaluated by Zimmermann [9] (see also [10] and [11]). Here, the dimensional regularization method was used [12-14]; in section (6), we want to show that the same result can be obtained with much less calculational effort using modern zeta-function techniques. This regularization scheme was originally developed by Dowker and Critchley [54] and by Hawking [15] to evaluate effective actions like (1.34) in curved space-time. Using a simple example, we also consider the case of finite temperature [18,19], which, as we shall see, possesses a formal analogy to the Casimir effect [20-23]. This is briefly shown in appendix C. Furthermore, in appendix B, we apply the same techniques to scalar QED.

Thereafter, in section (7), the two-loop calculation is performed with all renormalizations executed along the lines of conventional operator field theory. How the problem can be reformulated in the framework of Schwinger's Source Theory [29,30] is shown in appendix F.

In section (8), we come back to $L^{(1)}$, which is now considered from the viewpoint of renormalization group equations [4,12,31-35].

Finally, section (9) contains a further discussion of the QED vacuum structure as well as an outlook on some of the corresponding problems in quantum chromodynamics.

(2) The Electron Propagator in a Constant External Magnetic Field

In this section, we want to derive a special representation of the Dirac propagator of a particle in a constant external magnetic field. Since this point was already discussed in detail in [29] and [30], we shall simply sketch the course of calculation and compile the results which will be necessary later on.

The propagator, i.e., the causal Green's function, of a Dirac particle in an external field which is described by a potential

A^{μ} is defined by

$$G_{+} = \frac{1}{\gamma\Pi + m - i\varepsilon} \quad , \quad \varepsilon > 0 \tag{2.1}$$

with

$$\Pi_{\mu} = p_{\mu} - e A_{\mu} \tag{2.2}$$

where we have (and shall continue to) set ħ=c=1 and used the metric g=diag(-1,1,1,1). (See also Appendix A) From this representation of G_{+}, we get

$$G_{+} = \frac{\gamma\Pi - m}{(\gamma\Pi)^{2} - m^{2} + i\varepsilon} \tag{2.3}$$

Furthermore,

$$(\gamma\Pi)^{2} = -\Pi^{2} - \frac{i}{2}\sigma^{\mu\nu}[\Pi_{\mu}, \Pi_{\nu}] \tag{2.4}$$

where

$$\{\gamma^{\mu}, \gamma^{\nu}\} = -2g^{\mu\nu} \quad , \quad \frac{i}{2}[\gamma^{\mu}, \gamma^{\nu}] = \sigma^{\mu\nu} .$$

Since the commutator in (2.4) can be written as

$$[\Pi^{\mu}, \Pi^{\nu}] = i e F^{\mu\nu} \tag{2.5}$$

we obtain altogether

$$(\gamma\Pi)^{2} = -\Pi^{2} + \frac{e}{2}\sigma_{\mu\nu}F^{\mu\nu} \tag{2.6}$$

If one limits oneself to a constant magnetic field in z-direction, then only the components

$$F_{12} = -F_{21} \quad =: B = \text{const} \tag{2.7}$$

of the field strength tensor are non-vanishing, and we find

$$(\gamma\Pi)^{2} = -\Pi^{2} + eB\sigma^{12} = -\Pi^{2} + eB\sigma^{3} \tag{2.8}$$

Then the propagator can be written as

$$G_+ = \frac{m - \gamma\Pi}{\Pi^2 + \varkappa^2 - i\varepsilon} \tag{2.9}$$

with

$$\varkappa^2 := m^2 - eB\sigma^3 \tag{2.10}$$

Of special interest is the space representation of G_+:

$$G_+(x', x''|A) = \langle x'| \frac{m - \gamma\Pi}{\Pi^2 + \varkappa^2 - i\varepsilon} |x''\rangle \tag{2.11}$$

which satisfies the following Green's function equation:

$$[\gamma^\mu(\tfrac{1}{i}\partial'_\mu - eA_\mu(x')) + m]\, G_+(x', x''|A) = \delta(x' - x'') \tag{2.12}$$

To solve this equation we make the Ansatz

$$G_+(x', x''|A) = \Phi(x', x'')[m - \gamma^\nu(\tfrac{1}{i}\partial'_\nu - eA'_\nu(x'))]\,\Delta_+(x', x''|A') \tag{2.13}$$

with

$$\Phi(x', x'') := \exp\left[ie \int_{x''}^{x'} dx_\mu \{A^\mu(x) + \tfrac{1}{2}F^{\mu\nu}(x'_\nu - x''_\nu)\}\right] \tag{2.14}$$

and

$$A'^\mu(x') := -\tfrac{1}{2}F^{\mu\nu}(x' - x'')_\nu \tag{2.15}$$

just as in (2.7).

It is easy to show that the integral in (2.14) is independent of the choice of the path of integration, since the curl of the integrand vanishes. If one chooses a straight line as integration path,

$$x(t) = x'' + t(x' - x''), \quad t \in [0,1]$$

one finds that the second term of the integrand gives no contribution.

For a straight integration path, then

$$\Phi(x', x'') = \exp\left[ie \int_{x''}^{x'} dx_\mu A^\mu(x)\right] \tag{2.16}$$

In addition, we need the derivative of (2.14) with respect to the upper limit of the path integral:

$$\partial_\mu^{x'} \Phi(x', x'') = ie[A_\mu(x') - A'_\mu(x')]\,\Phi(x', x'') \tag{2.17}$$

If we then substitute the Ansatz (2.13) into (2.12) we obtain a differential equation for $\Delta_+ [A']$:

$$\left[\left(\tfrac{1}{i}\partial' - eA'\right)^2 + \varkappa^2\right] \Delta_+(x', x'' | A') = \delta(x' - x'') \quad (2.18)$$

With the definition of A', eq. (2.15), we then end up with the defining equation for $\Delta_+ [A']$:

$$\left[-\partial^2 + \varkappa^2 - i\varepsilon - \frac{e^2}{4} x_\mu F^{2\mu\nu} x_\nu\right] \Delta_+(x | A') = \delta(x) \quad (2.19)$$

Obviously, for a constant field the operator in the square bracket in (2.18) and (2.19) is translational invariant since it contains $A_\mu(x)$ only in a gauge invariant form as $F_{\mu\nu}$. Furthermore, use has been made of rotational invariance with respect to the z-axis which implies $F^{\mu\nu}(X_\mu \partial_\nu - X_\nu \partial_\mu)\Delta_+(X|A') = 0$. (See [29] for details). Finally, $F^{2\mu\nu}$ stands for $F^{\mu}{}_{\alpha} F^{\alpha\mu}$.

Equation (2.19) can be solved by a Fourier Ansatz [43]

$$\Delta_+(x | A') = \int \frac{d^4k}{(2\pi)^4}\, e^{ikx}\, \Delta_+(k | A') \quad (2.20)$$

thereby converting (2.19) into

$$\left[k^2 + \frac{e^2}{4} \partial_\mu^{(k)} F^{2\mu\nu} \partial_\nu^{(k)} + \varkappa^2 - i\varepsilon\right] \Delta_+(k | A') = 1 \quad (2.21)$$

Now we shall try to solve this equation by an Ansatz of the form

$$\Delta_+(k | A') = i \int_0^\infty ds\, e^{-M(is)}\, e^{-is(\varkappa^2 - i\varepsilon)} \quad (2.22)$$

with

$$M(is) = k^\alpha X_{\alpha\beta}(is) k^\beta + Y(is) \;, \quad X_{\alpha\beta} = X_{\beta\alpha} \quad (2.23)$$

Inserting (2.22) into (2.21) then yields

$$i \int_0^\infty ds \left[k \left\{1 + e^2 X(is) F^2 X(is)\right\} k - \frac{e^2}{2} \mathrm{tr}\left\{F^2 X\right\} + \varkappa^2 - i\varepsilon\right] \cdot$$

$$\cdot \exp\left\{-M(is)\right\} \exp\left\{-is(\varkappa^2 - i\varepsilon)\right\} = 1 \quad (2.24)$$

This equation for determining X and Y takes the form

$$i\int_0^\infty ds\; g(is)\, e^{-f(is)} = 1 \tag{2.25}$$

If we were now to set

$$g(is) = f'(is) \tag{2.26}$$

this would give

$$i\int_0^\infty ds\; g(is)\, e^{-f(is)} = e^{-f(0)} - \lim_{s\to\infty} e^{-f(is)} \tag{2.27}$$

that is, (2.25) is solved by our Ansatz if there are solutions of (2.26) with

$$f(0) = 0 \quad , \quad \lim_{s\to\infty} \operatorname{Re} f(is) = \infty . \tag{2.28}$$

The relation (2.26) reads in our case

$$k\,[1 + e^2 X(is)\, F^2 X(is)]\, k - \frac{e^2}{2}\operatorname{tr}(F^2 X(is)) = M'(is) \tag{2.29}$$

If we rotate the integration path according to is → s and we use (2.23), it follows that

$$\begin{aligned} 1 + e^2 X(s)\, F^2 X(s) &= \dot{X}(s) \\ -\frac{e^2}{2}\operatorname{tr}(F^2 X(s)) &= \dot{Y}(s) \end{aligned} \tag{2.30}$$

As one can easily prove by differentiation, these equations are solved by

$$\begin{aligned} X(s) &= (eF)^{-1}\tan(eFs) \\ Y(s) &= \frac{1}{2}\operatorname{tr}\ln\cos(eFs) \end{aligned} \tag{2.31}$$

Evidently X(0) = Y(0) = 0 is valid so that the first of the conditions in (2.28) is satisfied.

In order to further evaluate these solutions, we take advantage of the special form of the field strength tensor:

$$(F^{\mu\nu}) = \begin{pmatrix} 0 & 0 & 0 & 0 \\ 0 & 0 & B & 0 \\ 0 & -B & 0 & 0 \\ 0 & 0 & 0 & 0 \end{pmatrix} \quad , \quad (iF)^2 = B^2 \begin{pmatrix} 0 & & & 0 \\ & 1 & & \\ & & 1 & \\ 0 & & & 0 \end{pmatrix} \tag{2.32}$$

Then we find

$$X(is) = (eF)^{-1} \tan(ieFs) = is\left[\begin{pmatrix} 1 & & & \\ & 0 & & \\ & & 0 & \\ & & & 1 \end{pmatrix} + \begin{pmatrix} 0 & & & \\ & 1 & & \\ & & 1 & \\ & & & 0 \end{pmatrix} \frac{\tan(eBs)}{(eBs)}\right] \tag{2.33}$$

It is now convenient to introduce the notation

$$a_{\parallel} := (a^0, 0, 0, a^3) \quad , \quad a_{\perp} := (0, a^1, a^2, 0)$$

$$(ab)_{\parallel} = -a^0 b^0 + a^3 b^3 \quad , \quad (ab)_{\perp} := a^1 b^1 + a^2 b^2 \tag{2.34}$$

for arbitrary Lorentz vectors a, b.

Then

$$k X(is) k = is\left[k_{\parallel}^2 + \frac{\tan(eBs)}{eBs} k_{\perp}^2\right] \tag{2.35}$$

(Note that $X \equiv (X_{\alpha}{}^{\beta})$.) Accordingly we get

$$\begin{aligned} Y(is) &= \tfrac{1}{2} \operatorname{tr} \ln \cos(ieFs) \\ &= \tfrac{1}{2} \ln \det \left[\begin{pmatrix} 1 & & & \\ & 0 & & \\ & & 0 & \\ & & & 1 \end{pmatrix} + \begin{pmatrix} 0 & & & \\ & 1 & & \\ & & 1 & \\ & & & 0 \end{pmatrix} \cos(eBs)\right] \\ &= \ln \cos(eBs) \end{aligned} \tag{2.36}$$

Let us recall the definition

$$f(is) = M(is) + is(\varkappa^2 - i\varepsilon) = kX(is)k + Y(is) + is(\varkappa^2 - i\varepsilon)$$

Thus, it becomes clear that the solutions (2.31) also fulfill the second condition (2.28):

$$\lim_{s\to\infty} e^{-f(is)} = \lim_{s\to\infty} \exp\left\{-is\left(k_{\parallel}^2 + k_{\perp}^2 \frac{\tan(eBs)}{eBs}\right)\right\} \frac{e^{-is(\varkappa^2 - i\varepsilon)}}{\cos(eBs)} = 0 \tag{2.37}$$

From (2.20), (2.23) and (2.31) it now follows for the scalar Green's function Δ_+:

$$\Delta_+(x',x''|A) = i\int_0^\infty ds\ \phi(x',x'')\int\frac{d^4k}{(2\pi)^4}\,e^{ik(x'-x'')}\cdot$$
$$\cdot\frac{1}{\cos z}\exp\left\{-is\left(k_\parallel^2+k_\perp^2\frac{\tan z}{z}\right)\right\}e^{-is(\varkappa^2-i\varepsilon)} \tag{2.38}$$

with (2.39)

$$z := eBs$$

But beginning directly with the propagator

$$\Delta_+(x',x''|A) = \langle x'|\frac{1}{\pi^2+\varkappa^2-i\varepsilon}|x''\rangle$$
$$= i\int_0^\infty ds\ \langle x'|e^{-is\pi^2}|x''\rangle\, e^{-is(\varkappa^2-i\varepsilon)} \tag{2.40}$$

a comparison with (2.38) shows that

$$\langle x'|e^{-is\pi^2}|x''\rangle = \phi(x',x'')\int\frac{d^4k}{(2\pi)^4}e^{ik(x'-x'')}\frac{1}{\cos z}\,e^{-is\left(k_\parallel^2+k_\perp^2\frac{\tan z}{z}\right)} \tag{2.41}$$

From this important formula, further useful relations can be derived; let us substitute $s \to a_\perp s$; then

$$\langle x'|e^{isa_\perp\pi^2}|x''\rangle = \phi(x',x'')\int\frac{d^4k}{(2\pi)^4}\,e^{ik(x'-x'')}\,\frac{1}{\cos(a_\perp z)}\cdot$$
$$\cdot\exp\left\{-is\left(k_\parallel^2+k_\perp^2\,\frac{\tan(a_\perp z)}{(a_\perp z)}\right)a_\perp\right\} \tag{2.42}$$

Because

$$\langle x'|e^{-isa_\perp\pi^2}\pi_\mu|x''\rangle = (i\partial''-eA(x''))_\mu\,\langle x'|e^{-isa_\perp\pi^2}|x''\rangle \tag{2.43}$$

we get

$$\langle x'|\exp\{-is[a^{(0)}\pi_0\pi^0+a^{(3)}\pi_3\pi^3+a_\perp\pi_\perp^2]\}|x''\rangle$$
$$=\langle x'|e^{-isa_\perp\pi^2}\exp\{-is[(a^{(0)}-a_\perp)\pi_0\pi^0+(a^{(3)}-a_\perp)\pi_3\pi^3]\}|x''\rangle \tag{2.44}$$
$$=\exp\{-is[-(a^{(0)}-a_\perp)(i\partial^{0''}-eA^0(x''))^2+(a^{(3)}-a_\perp)(i\partial^{3''}-eA^3(x''))^2]\}\cdot$$
$$\cdot\langle x'|e^{-isa_\perp\pi^2}|x''\rangle$$

With
$$\partial^{\mu''}\phi(x',x'') = -ie\left[A^{\mu}(x'') - \tfrac{1}{2}F^{\mu\nu}(x'-x'')_{\nu}\right]\phi(x',x'')$$
$$F^{0\mu} = F^{3\mu} = 0$$
and
$$\exp\left\{-is\left[-(a^{(0)}-a_{\perp})(i\partial^{0''})^{2} + (a^{(3)}-a_{\perp})(i\partial^{3''})^{2}\right]\right\}\cdot$$
$$\cdot\, e^{ik(x'-x'')} = \exp\left\{-is\left[-(a^{(0)}-a_{\perp})(k^{0})^{2} + (a^{(3)}-a_{\perp})(k^{3})^{2}\right]\right\} e^{ik(x'-x'')}$$
it follows from (2.43) and (2.44)
$$\langle x'|\exp\{-is[a^{(0)}\Pi_0\Pi^0 + a^{(3)}\Pi_3\Pi^3 + a_{\perp}\Pi_{\perp}^2]\}|x''\rangle$$
$$= \phi(x',x'')\int\frac{d^4k}{(2\pi)^4}\, e^{ik(x'-x'')}\frac{1}{\cos(a_{\perp}z)}\exp\Big\{-is\Big[a^{(0)}k_0k^0 + \tag{2.45}$$
$$+ a^{(3)}k_3k^3 + a_{\perp}k_{\perp}^2\,\frac{\tan(a_{\perp}z)}{(a_{\perp}z)}\Big]\Big\}$$

Furthermore, we need matrixelements of the following form
$$\langle x'|\exp\{-is[a^{(0)}\Pi_0\Pi^0 + a^{(3)}\Pi_3\Pi^3 + a_{\perp}\Pi_{\perp}^2\}\,\Pi^{\mu}|x''\rangle =$$
$$= \phi(x',x'')\int\frac{d^4k}{(2\pi)^4}\, e^{ik(x'-x'')}\left[k^{\mu} - ea_{\perp}sF^{\mu\nu}k_{\nu}\frac{\tan(a_{\perp}z)}{(a_{\perp}z)}\right]\frac{1}{\cos(a_{\perp}z)}\cdot$$
$$\cdot\exp\Big\{-is\Big[a^{(0)}k_0k^0 + a^{(3)}k_3k^3 + a_{\perp}k_{\perp}^2\,\frac{\tan(a_{\perp}z)}{(a_{\perp}z)}\Big]\Big\}$$
From this equation, together with
$$\gamma_{\mu}F^{\mu\nu}k_{\nu}\tan(a_{\perp}z) = (\gamma_1k_2 - \gamma_2k_1)\,B\tan(a_{\perp}z)$$
$$= \frac{i}{\cos(a_{\perp}z)}\,B\sin(a_{\perp}z\sigma^3)(\gamma k)_{\perp}$$

and
$$(k\gamma)_{\perp} - esa_{\perp}(\gamma_{\mu}F^{\mu\nu}k_{\nu})\frac{\tan(a_{\perp}z)}{(a_{\perp}z)}$$
$$= \frac{1}{\cos(a_{\perp}z)}\,e^{-ia_{\perp}z\sigma^3}(\gamma k)_{\perp}$$

follow the important relations

$$\langle x'|\exp\{-is[a^{(0)}\Pi_0\Pi^0 + a^{(3)}\Pi_3\Pi^3 + a_\perp \Pi_\perp^2]\}(1, \gamma^0\Pi^0, \gamma^3\Pi^3, (\gamma\Pi)_\perp)|x''\rangle$$

$$= \phi(x',x'')\int\frac{d^4k}{(2\pi)^4}e^{ik(x'-x'')}\frac{1}{\cos(eBsa_\perp)}\cdot$$

$$\cdot\exp\{-is[a^{(0)}k_0k^0 + a^{(3)}k_3k^3 + a_\perp\frac{\tan(eBsa_\perp)}{(eBsa_\perp)}k_\perp^2]\}\cdot \quad (2.46)$$

$$\cdot\left(1, \gamma_0 k^0, \gamma_3 k^3, \frac{1}{\cos(ea_\perp Bs)}e^{-ieBsa_\perp\sigma^3}(\gamma k)_\perp\right)$$

For the propagator

$$G_+(x',x'') = \langle x'|\frac{m-\gamma\Pi}{\Pi^2+\varkappa^2-i\varepsilon}|x''\rangle$$

$$= i\int_0^\infty ds\, e^{-is(\varkappa^2-i\varepsilon)}\langle x'|e^{-is\Pi^2}(m-\gamma\Pi)|x''\rangle$$

we thus get

$$G_+(x',x'') = \phi(x',x'')\int\frac{d^4k}{(2\pi)^4}e^{ik(x'-x'')}g(k) \quad (2.47a)$$

where

$$g(k) = i\int_0^\infty ds\,\exp\{-is[m^2-i\varepsilon+k_\parallel^2+\frac{\tan z}{z}k_\perp^2]\}\cdot$$

$$\cdot\frac{e^{i\sigma^3 z}}{\cos z}\left(m-\gamma k_\parallel - \frac{e^{-i\sigma^3 z}}{\cos z}\gamma k_\perp\right) \quad (2.47b)$$

We shall use g(k) in this form in the following sections, in order to calculate the mass operator and the polarization tensor; furthermore, formula (2.46) allows us to easily transform the thus calculated mass operator from the momentum into the position space representation.

(3) The Mass Operator in a Constant External Magnetic Field

In deriving the electron propagator in the last section, we did not take into account any radiative corrections, i.e., G_+, according to (2.47) can be considered as the first term of an expansion of the complete two-point function, i.e., of the dressed propagator

$$G'_+(x',x'') = i\langle 0|T\,\psi(x')\,\bar{\psi}(x'')|0\rangle \tag{3.1}$$

in powers of the coupling constant α. For now, if we consider the case of a vanishing external field, then $G'_+(x',x'') = G'_+(x'-x'')$ is translation invariant and we can define its Fourier transform by

$$G'_+(p) = \int d^4x\; e^{-ip\cdot x}\; G'_+(x)$$

The first terms of the perturbation series for $G'_+(p)$ are presented by the following Feynman graphs:

$G'_+(p) =$

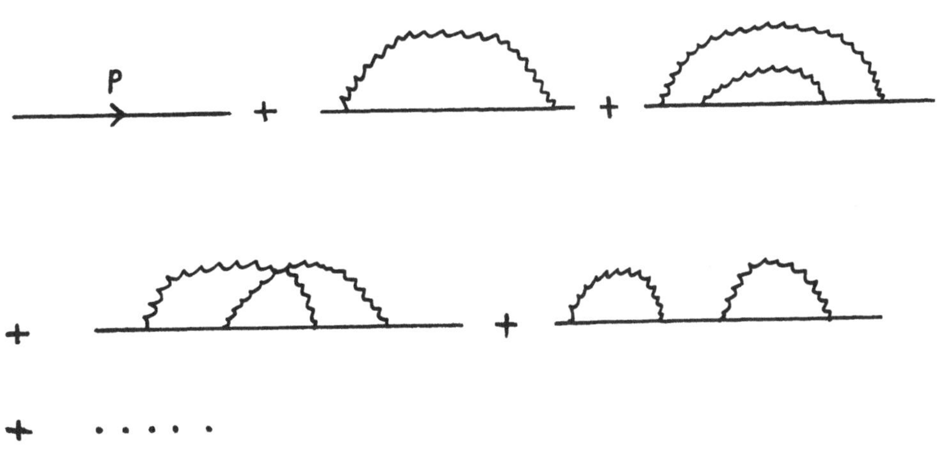

The latter diagram is thereby one-particle-reducible, since it can be divided into two diagrams by cutting <u>one</u> inner propagator, i.e., such a reducible diagram can be obtained through iteration of simpler graphs. It is therefore useful to introduce the "proper" self-energy part of the electron or mass operator $\Sigma(p)$, by the sum of all contributing irreducible graphs. Iteration of Σ then gives

$$G_+'(p) = \frac{1}{\not{p} + m_0 + \Sigma(p)}$$

where m_0 represents the unrenormalized (bare) mass of the electron. If we limit ourselves now to the first order in α, then only the diagram

has to be calculated, which, according to the usual Feynman rules, leads to

$$\Sigma^{(2)}(p) = ie^2 \gamma^\mu \int \frac{d^4k}{(2\pi)^4} \frac{1}{k^2 - i\varepsilon} G_+(p-k)\, \gamma_\mu \quad .$$

where we chose the Feynman gauge for the photon propagator: $D_{+\mu\nu} = g_{\mu\nu} D_+$. The superscript (2) in $\Sigma^{(2)}$ indicates the second-order approximation in the coupling constant e of the entire mass operator Σ. We now wish to determine what influence the consideration of radiative corrections has on the pole structure of $G_+'^{(2)}(p)$. For this purpose, we define a mass m by

$$\left\{ \not{p} + m_0 + \Sigma^{(2)}(\not{p}) \right\} \Big|_{\not{p} = -m} = 0$$

i.e., as the pole of the propagator $G_+^{\prime(2)}(p)$ in the presence of interactions with the photon field. Now we expand the mass operator according to

$$\Sigma^{(2)}(\not p) = A + B\cdot(\not p + m) + C(\not p)$$

with constants

$$A \equiv \Sigma(\not p)\Big|_{\not p=-m} \, , \qquad B \equiv \frac{\partial \Sigma(\not p)}{\partial \not p}\Bigg|_{\not p=-m}$$

and $C(\not p)$ = quadratic and higher order terms in $(\not p+m)$.

In terms of these quantities, $G_+^{\prime(2)}(p)$ yields

$$G_+^{\prime(2)}(p) = \frac{1}{\not p + m_0 + \delta m + B\cdot(\not p + m) + C(\not p)}$$

where we have used the customary notation $A \equiv \delta m$.
Near the mass shell, i.e., for $\not p \sim -m$ or $p^2+m^2 \approx 0$, this reduces to

$$G_+^{\prime(2)}(\not p \approx -m) = \frac{1}{\not p + m_0 + \delta m}$$

from which we can conclude, because of the defining equation for m as pole of $G_+^{\prime(2)}(p)$ on the mass shell, that the bare mass m_0 is associated with the physical (renormalized) mass m by

$$m = m_0 + \delta m \, , \qquad \delta m = \Sigma^{(2)}(\not p)\Big|_{\not p=-m}$$

The fact that we are confronted with a non-vanishing mass operator $\Sigma^{(2)}(p)|_{\not p=-m}$ when considering interactions between the electromagnetic field and the Dirac field forces us to redefine the mass parameter of the theory according to the last equation, so that the pole of the electron propagator is given by the physical (observed) mass m. It should be noted that the necessity of such a renormalization is independent of whether or not the quantity δm is divergent. Thus we obtain

$$G_+^{1(2)}(p) = \frac{1}{\not p + m + B\cdot(\not p + m) + C(\not p)}$$

$$= \frac{(1+B)^{-1}}{\not p + m + (1+B)^{-1} C(\not p)}$$

Near the mass shell, C approaches zero quadratically so that we obtain

$$G_+^{1(2)}(\not p \approx -m) = \frac{Z_2}{\not p + m} \quad , \quad Z_2 \equiv (1+B)^{-1}$$

In order for the normalization of the electron wave function to be retained [39], the residue of the pole at $\not p = -m$ must be one, which is not, however, the case for $B \neq 0$. This leads us to interpret the fields appearing in (3.1) as bare wave functions ψ_0, $\bar\psi_0$, which according to

$$\psi = Z_2^{-1/2} \psi_0$$

yield renormalized fields ψ , $\bar\psi$ which give us a pole in the electron propagator with residue one. The necessity of such a renormalization is independent of whether Z_2 is divergent or not; in QED, though, both δm and Z_2 are established to be divergent quantities.

Finally, for the renormalized dressed electron propagator, we obtain perturbatively, to order e^2:

$$G'^{(2)}_{+\,ren}(p) = \frac{G'^{(2)}_{+}(p)}{Z_2}$$

$$= \frac{1}{\not p + m + (1+B)^{-1} C(\not p)} = \frac{1}{\not p + m + C(\not p)} + O(e^4)$$

where we took advantage of the fact that both $C(\not p)$ and B are of order e^2, i.e., up to this order, the term $B \cdot C(\not p)$ can be neglected. If we compare the last expression with the original equation for the unrenormalized electron propagator $G'_+(p) = (\not p + m_0 + \Sigma(\not p))^{-1}$, we conclude that in the process of renormalization, the bare mass m_0 has been replaced by the physical mass m and the function $C(\not p)$ has taken the place of the unrenormalized mass operator $\Sigma^{(2)}$ where C agrees with $\Sigma^{(2)}$ from the quadratic term in $(\not p + m)$ on, but does not contain constant and linear terms. Therefore, to calculate the renormalized mass operator $\Sigma^{(2)}_{Ren}(\not p) = C(\not p)$ (which we shall again refer to by $\Sigma(p)$ in the following), we can also proceed by starting with the ansatz

$$\Sigma^{(2)}_{Ren}(p) \equiv \Sigma(p) = ie^2 \gamma_\mu \int \frac{d^4k}{(2\pi)^4} \frac{1}{k^2 - i\varepsilon} G_+(p-k) \gamma^\mu + c.t.$$

and choosing the counter terms c.t. in such a manner that

$$\Sigma(\not p = -m) = 0 \;, \quad \frac{\partial \Sigma}{\partial \not p}(\not p = -m) = 0$$

leading to A = B = 0. $G_+(p-k)$ is thereby parametrized by the physical mass m.

Following Tsai [7], we now want to calculate the diagram

$\Sigma =$  (3.4)

explicitly, whereby we shall use the electron propagator in the external field indicated by the double line in (3.4). In space representation, (3.4) leads to the expression

$$\Sigma(x',x'') = ie^2 \gamma^\mu G_+(x',x'') D_+(x'-x'') \gamma_\mu + c.t. \tag{3.5}$$

with the free photon propagator

$$D_+(x) = \int \frac{d^4k}{(2\pi)^4} e^{ikx} \frac{1}{k^2 - i\varepsilon} \tag{3.6}$$

and

$$G_+(x',x'') = \Phi(x',x'') \int \frac{d^4p}{(2\pi)^4} e^{ip(x'-x'')} g(p) \tag{3.7}$$

where $g(p)$ is given by (2.47b).

Substituion gives

$$\Sigma(x',x'') = \Phi(x',x'') \int \frac{d^4p}{(2\pi)^4} e^{ip(x'-x'')} \Sigma(p) \tag{3.8}$$

with

$$\Sigma(p) = i e^2 \gamma^\mu \int \frac{d^4k}{(2\pi)^4} \frac{1}{k^2 - i\varepsilon} g(p-k)\gamma_\mu + c.t. \tag{3.9}$$

Using (2.47b) and the proper time integral

$$\frac{1}{k^2 - i\varepsilon} = i \int_0^\infty ds_2 \, e^{-is_2(k^2 - i\varepsilon)} \tag{3.10}$$

we obtain

$$\Sigma(p) = i e^2 \gamma^\mu \int \frac{d^4k}{(2\pi)^4} \left\{ i \int_0^\infty ds_2 \, e^{-is_2(k^2-i\varepsilon)} \right\} \cdot$$

$$\cdot \left\{ i \int_0^\infty ds_1 \, e^{-is_1\left(m^2 + (p-k)_\parallel^2 + \frac{\tan z}{z}(p-k)_\perp^2\right)} \right.$$

$$\left. \cdot \frac{e^{i\sigma^3 z}}{\cos z} \left[m - \gamma(p-k)_\parallel - \frac{e^{-i\sigma^3 z}}{\cos z} \gamma(p-k)_\perp \right] \right\} \gamma_\mu + c.t.$$

$$= -i e^2 \int_0^\infty ds_1 \int_0^\infty ds_2 \int \frac{d^4k}{(2\pi)^4} e^{-is_2(k^2 - i\varepsilon)}$$

$$\cdot \, e^{-is_1\left(m^2 + (p-k)_\parallel^2 + \frac{\tan z}{z}(p-k)_\perp^2\right)} \tag{3.11}$$

$$\cdot \, \gamma^\mu \frac{e^{i\sigma^3 z}}{\cos z} \left[m - \gamma(p-k)_\parallel - \frac{e^{-i\sigma^3 z}}{\cos z} \gamma(p-k)_\perp \right] \gamma_\mu + c.t.$$

Now we perform a transformation of variables for the s_1, s_2 integration:

$$s_1 =: su$$

$$s_2 =: s(1-u)$$

$$\int_0^\infty ds_1 \int_0^\infty ds_2 \ldots = \int_0^\infty ds\, s \int_0^1 du \ldots \tag{3.12}$$

$$z = eBs_1 \longrightarrow eBsu =: y$$

The exponential functions in (3.11) can then be rearranged as follows:

$$e^{-is_2k^2} \exp\left[-is_1\left(m^2 + (p-k)_\parallel^2 + \frac{\tan z}{z}(p-k)_\perp^2\right)\right]$$

$$= \exp\Big(-is\Big[(1-u)k^2 + u\Big\{m^2 + p_\parallel^2 - 2p\cdot k_\parallel + k_\parallel^2 + \frac{\tan y}{y}\cdot$$

$$\cdot (p_\perp^2 - 2p\cdot k_\perp + k_\perp^2)\Big\}\Big]\Big) \tag{3.13}$$

$$= e^{-is\chi}$$

with

$$\chi := um^2 + \varphi + (k-up)_\parallel^2 + \left(1-u+u\frac{\tan y}{y}\right)\left[k - \frac{u \tan y/y}{1-u+u\tan y/y}p\right]_\perp^2 \tag{3.14}$$

and

$$\varphi := u(1-u)p_\parallel^2 + \frac{u}{y}\frac{(1-u)\sin y}{(1-u)\cos y + u\sin y/y}p_\perp^2 \tag{3.15}$$

With these rearrangements it follows from (3.11)

$$\Sigma(p) = -ie^2 \int_0^\infty s\,ds \int_0^1 du \frac{1}{\cos y} \int \frac{d^4k}{(2\pi)^4} e^{-is\chi(u,k)} \cdot$$

$$\cdot \gamma^\mu e^{i\sigma^3 y}\left[m - \gamma(p-k)_\parallel - \frac{e^{-i\sigma^3 y}}{\cos y}\gamma(p-k)_\perp\right]\gamma_\mu + \text{c.t.} \tag{3.16}$$

Further simplification yields

$$\Sigma(p) = -ie^2 \int_0^\infty s\,ds \int_0^1 du \frac{1}{\cos Y} \left\{ \int \frac{d^4k}{(2\pi)^4} e^{-is\chi} \right\} \cdot$$

$$\cdot \gamma^\mu e^{i\sigma^3 Y} \left[m - (1-u)\gamma p_\parallel + \frac{e^{-i\sigma^3 Y}}{\cos Y} \frac{1-u}{1-u+u\tan Y/Y} \gamma p_\perp \right] \gamma_\mu + \text{c.t.} \tag{3.17}$$

The remaining k-integration is simple to do; for this, we first need the formulae [36]

$$\int_{-\infty}^{\infty} dx \quad \cos(Ax^2) = \left(\frac{\pi}{2A}\right)^{\frac{1}{2}} , \quad A>0$$

$$\int_{-\infty}^{\infty} dx \quad \sin(Ax^2) = \left(\frac{\pi}{2A}\right)^{\frac{1}{2}} , \quad A>0 \tag{3.18}$$

Then

$$\int_{-\infty}^{\infty} dx \quad e^{\pm iAx^2} = e^{\pm i\frac{\pi}{4}} \left(\frac{\pi}{A}\right)^{\frac{1}{2}} , \quad A>0 \tag{3.19}$$

With

$$A := 1 - u + u \tan Y/Y \tag{3.20}$$

we get, by shifting the integration variables

$$\int \frac{d^4k}{(2\pi)^4} e^{-is\chi} = \int \frac{d^4k}{(2\pi)^4} \exp\left[-is \left\{ \varphi + um^2 + (k-up)_\parallel^2 + A\left[k - \frac{u\tan Y/Y}{1-u+u\tan Y/Y} p\right]_\perp^2 \right\} \right] \tag{3.21}$$

$$= \frac{(-i)}{(4\pi)^2} \frac{1}{s^2} e^{-is(\varphi + um^2)} \frac{1}{1+u+u\tan Y/Y}$$

So far we obtain for the mass operator

$$\Sigma(p) = \frac{(-i)^2 e^2 m}{(4\pi)^2} \int_0^\infty \frac{ds}{s} \int_0^1 du \frac{e^{-is(um^2+\varphi)}}{(1-u)\cos Y + u\sin Y/Y} \cdot \tag{3.22}$$

$$\cdot [A_1 + A_2 + A_3] + \text{c.t.}$$

using the abbreviations

$$A_1 = \gamma^\mu e^{i\sigma^3 y} \gamma_\mu$$

$$A_2 = \gamma^\mu e^{i\sigma^3 y} \left[-(1-u) \frac{\gamma p_\parallel}{m}\right] \gamma_\mu \tag{3.23}$$

$$A_3 = -\frac{1-u}{(1-u)\cos y + u \sin y/y} \gamma^\mu \frac{\gamma p_\perp}{m} \gamma_\mu$$

With the aid of the formulae given in Appendix A, the Dirac algebra of the A's can be further simplified.

$$A_1 = -2 e^{i\sigma^3 y} \left[1 + e^{-2i\sigma^3 y}\right] \tag{3.24}$$

The next term yields, accordingly

$$A_2 = -2 e^{i\sigma^3 y} \left[(1-u) \frac{\gamma p_\parallel}{m} e^{-2i\sigma^3 y}\right] \tag{3.25}$$

The last term A_3 can be written in the form

$$A_3 = -2 e^{i\sigma^3 y} \left[\frac{1-u}{(1-u)\cos y + u \sin y/y} \frac{\gamma p_\perp}{m} e^{-i\sigma^3 y}\right] \tag{3.26}$$

If we put the A's into (3.22), it follows that

$$\Sigma(p) = \frac{\alpha m}{2\pi} \int_0^\infty \frac{ds}{s} \int_0^1 du \frac{e^{-is(um^2+\varphi)}}{(1-u)\cos y + u \sin y/y} e^{i\sigma^3 y} \cdot$$

$$\cdot \left[1 + e^{-2i\sigma^3 y} + (1-u) e^{-2i\sigma^3 y} \frac{\gamma p_\parallel}{m} + (1-u) \frac{e^{-i\sigma^3 y}}{(1-u)\cos y + u \sin y/y} \cdot \frac{\gamma p_\perp}{m}\right] + \text{c.t.} \tag{3.27}$$

with the fine structure constant $\alpha = e^2/(4\pi)$.

The remaining integrations cannot be carried out in a closed form; it is, however, easy with the help of the transformation amplitudes (2.46) to convert (3.27) into a space representation. For this we use the two auxiliary equations

$$\phi(x',x'')\int\frac{d^4p}{(2\pi)^4}\,e^{ip(x'-x'')}e^{-is\varphi} = \phi(x',x'')\int\frac{d^4p}{(2\pi)^4}\,e^{ip(x'-x'')}\cdot$$

$$\cdot\, e^{-isu(1-u)p_{\parallel}^2}\exp\left[-\frac{i}{eB}\left\{\frac{(1-u)\sin y}{(1-u)\cos y+u\sin y/y}\right\}p_{\perp}^2\right]$$

$$= \cos\beta\ \langle x'|\,e^{-isu(1-u)p_{\parallel}^2}\,e^{-i\frac{\beta}{eH}\pi_{\perp}^2}\,|x''\rangle \tag{3.28}$$

and

$$\phi(x',x'')\int\frac{d^4p}{(2\pi)^4}\,e^{ip(x'-x'')}e^{-is\varphi}(a\,\gamma p_{\parallel}+b\,\gamma p_{\perp}) = \tag{3.29}$$

$$\cos\beta\,\langle x'|\,e^{-isu(1-u)p_{\parallel}^2}\,e^{-i\frac{\beta}{eB}\pi_{\perp}^2}(a\,\gamma p_{\parallel}+b\cos\beta\, e^{i\sigma^3\beta}\,\gamma\pi_{\perp})|x''\rangle$$

where we set

$$\tan\beta := \frac{(1-u)\sin y}{(1-u)\cos y+u\sin y/y} \tag{3.30}$$

From this, we get for cos β

$$\cos\beta = (1+\tan^2\beta)^{\frac{1}{2}}$$

$$= \left[1+\frac{(1-u)^2\sin^2 y}{[(1-u)\cos y+u\sin y/y]^2}\right]^{\frac{1}{2}} \tag{3.31}$$

$$= \{(1-u)\cos y+u\sin y/y\}\ \Delta^{-\frac{1}{2}}$$

in abbreviated form

$$\Delta := (1-u)^2+2u(1-u)\cos y\,\sin y/y + u^2\sin^2 y/y^2$$

With (3.28) and (3.29) the mass operator takes the space representation

$$\Sigma(x',x'') = \phi(x',x'') \int \frac{d^4p}{(2\pi)^4}\, e^{-ip(x'-x'')}\, \Sigma(p)$$

$$= \langle x' | \Sigma(\pi) | x'' \rangle \tag{3.32}$$

with

$$\Sigma(\pi) = \frac{\alpha m}{2\pi} \int_0^\infty \frac{ds}{s} \int_0^1 du\, e^{-ism^2u} \{C_1 + C_2 + C_3\} \tag{3.33}$$

C_1 denotes the contribution of the first two terms in the parenthesis of (3.27), i.e., of

$$\left\{\phi(x',x'') \int \frac{d^4p}{(2\pi)^4}\, e^{ip(x'-x'')} e^{-is\varphi}\right\} e^{i\sigma^3 y}\, \frac{1+e^{-2i\sigma^3 y}}{(1+u)\cos y + u \sin y/y}$$

to $\Sigma(\Pi)$. Now we can use (3.28) with (3.31) and get

$$C_1 = \{(1-u)\cos y + u \sin y/y\}\, \Delta^{-\frac{1}{2}}\, e^{-isu(1-u)p_\parallel^2} \cdot$$

$$\cdot\, e^{-i\frac{\beta}{eB}\pi_\perp^2} \cdot \frac{e^{i\sigma^3 y}(1+e^{-2i\sigma^3 y})}{(1-u)\cos y + u \sin y/y} \tag{3.34}$$

$$= \Delta^{-\frac{1}{2}}\, e^{-is\phi}\, e^{isum^2 - isu^2m^2}\, (1+e^{-2i\sigma^3 y})$$

with

$$\phi := u(1-u)[m^2 - \Pi^2] + \frac{u}{y}[\beta-(1-u)y]\pi_\perp^2 - u^2 \frac{e}{2}\sigma_{\mu\nu}F^{\mu\nu}, \tag{3.35}$$

since from this definition and

$$-\Pi^2 = \pi^2 - \frac{e}{2}\sigma_{\mu\nu}F^{\mu\nu}, \quad \frac{1}{2}\sigma_{\mu\nu}F^{\mu\nu} = B\sigma^3$$

the first line of (3.34) again follows.

Similarly we get for the contribution of the third term in the parenthesis of (3.27), because of (3.29)

$$C_2 = \Delta^{-\frac{1}{2}} e^{-is\phi} e^{isum^2 - isu^2m^2} (1-u) e^{-2i\sigma^3 y} \frac{\gamma\Pi_{\parallel}}{m} \quad (3.36)$$

Finally, we must evaluate C_3:

$$C_3 = \frac{(1-u)e^{-i\sigma^3 y}}{(1-u)\cos y + u \sin y/y} e^{i\sigma^3\beta} \{(1-u)\cos y + u \sin y/y\} \Delta^{-\frac{1}{2}} \cdot \frac{\gamma\Pi_{\perp}}{m} \cdot$$

$$\cdot \Delta^{-\frac{1}{2}} e^{-is\phi} e^{isum^2 - isu^2m^2}$$

First, we need

$$\sin\beta = (1-\cos^2\beta)^{\frac{1}{2}}$$

$$\underset{(3.30)}{=} \left[1 - \Delta^{-1}\{(1-u)^2\cos^2 y + 2u(1-u)\sin y \cos y/y + u^2 \sin^2 y/y^2\}\right]^{\frac{1}{2}}$$

$$= \Delta^{-\frac{1}{2}} (1-u) \sin y$$

from which we obtain

$$e^{i\sigma^3\beta} = \cos\beta + i\sigma^3 \sin\beta$$

$$= \Delta^{-\frac{1}{2}}\{(1-u)\cos y + u \sin y/y\} + i\sigma^3 \Delta^{-\frac{1}{2}} (1-u) \sin y$$

$$= \Delta^{-\frac{1}{2}} (1-u) e^{i\sigma^3 y} + \Delta^{-\frac{1}{2}} u \sin y/y$$

We can now substitute this into the above expression for C_3:

$$C_3 = \Delta^{-\frac{1}{2}} (1-u) e^{-i\sigma^3 y} \left[\frac{1-u}{\Delta} e^{i\sigma^3 y} + \frac{u}{\Delta} \frac{\sin y}{y}\right] e^{-is\phi} e^{isum^2 - isu^2m^2}$$

$$= \Delta^{-\frac{1}{2}} e^{-is\phi} e^{isum^2 - isu^2m^2} \left[(1-u)\left(\frac{1-u}{\Delta} + \right.\right. \quad (3.37)$$

$$\left. + \frac{u}{\Delta} \frac{\sin y}{y} e^{-i\sigma^3 y}\right]$$

With these representations of C_i, we get from (3.32) as end result for the unrenormalized mass operator

$$\Sigma(\pi) = \frac{\alpha m}{2\pi}\int_0^\infty \frac{ds}{s}\int_0^1 du\, e^{-isu^2m^2}\Big\{\Delta^{-\frac{1}{2}} e^{-is\phi}\,\Big[1+$$

$$+ e^{-2i\sigma^3 y} + (1-u)\,e^{-2i\sigma^3 y}\frac{\gamma\pi}{m} + (1-u)\Big(\frac{1-u}{\Delta} +$$

$$+\frac{u}{\Delta}\frac{\sin y}{y}\,e^{-i\sigma^3 y} - e^{-2i\sigma^3 y}\Big)\frac{\gamma\pi_\perp}{m}\Big] + \mathrm{c.t.}\Big\} \tag{3.38a}$$

with

$$\phi = u(1-u)\left[m^2-(\gamma\pi)^2\right] + \frac{u}{y}\left[\beta+(1-u)y\right]\pi_\perp^2 - u^2\frac{e}{2}\sigma_{\mu\nu}F^{\mu\nu} \tag{3.38b}$$

and

$$\Delta = (1-u)^2 + 2u(1-u)\frac{\sin y\,\cos y}{y} + u^2\left(\frac{\sin y}{y}\right)^2 \tag{3.38c}$$

The counter terms c.t. are now to be so determined that, after turning off the field, the mass operator renormalization conditions are fulfilled, i.e., that

$$\lim_{\gamma\pi\to -m}\ \lim_{B\to 0}\ \Sigma = 0 \tag{3.39}$$

and

$$\lim_{\gamma\pi\to -m}\ \lim_{B\to 0}\ \frac{\partial\Sigma}{\partial \not\pi} = 0 \tag{3.40}$$

is valid. First in the limiting case $y = eBsu \to 0$:

$$\beta = \arctan\left[\frac{(1-u)\sin y}{(1-u)\cos y + u\sin y/y}\right]$$

$$= \arctan\left[(1-u)y + O(y^2)\right]$$

$$= (1-u)y + O(y^2)$$

just as

$$\phi(B=0) = u(1-u)[m^2 - \not{\Pi}^2] + u\,\underbrace{\left[\frac{\beta}{y} - (1-u)\right]}_{\to 0}\Pi_\perp^2 - u^2\frac{e^2}{2}\underbrace{\sigma_{\mu\nu}F^{\mu\nu}}_{\to 0}$$

$$= u(1-u)[m^2 - \not{\Pi}^2]$$

Furthermore, it follows that

$$\Delta(B=0) = (1-u)^2 + 2u(1-u)\underbrace{\frac{\sin y}{y}\cos y}_{\to 1} + u^2\left(\frac{\sin y}{y}\right)^2 = 1$$

From (3.38) we thus get the mass operator $\Sigma(B=0) := \Sigma_0$ for the field-free case

$$\Sigma_0 = \frac{\alpha m}{2\pi}\int_0^\infty \frac{ds}{s}\int_0^1 du\, e^{-isu^2m^2}\left\{e^{-isu(1-u)}(m^2 - p^2)\cdot\right. \tag{3.41}$$

$$\left.\cdot\left[2 + (1-u)\frac{p}{m}\right] + \text{c.t.}\right\}$$

So the value of Σ_0 (without counterterms) on the mass shell is

$$\Sigma_0(p = -m) = \frac{\alpha m}{2\pi}\int_0^\infty \frac{ds}{s}\int_0^1 du\, e^{-isu^2m^2}\left[2 + (1-u)\frac{(-m)}{m}\right]$$

$$= \frac{\alpha m}{2\pi}\int_0^\infty \frac{ds}{s}\int_0^1 du\, e^{-isu^2m^2}(1+u) \tag{3.42}$$

But according to (3.39), this should vanish, i.e., we must choose the first part of the counter terms in the form

$$\text{c.t.}^{(1)} = -(1+u) \tag{3.43a}$$

so that $\Sigma_0(p = -m) = 0$. To fulfill (3.40), we need the derivative

$$\frac{\partial\Sigma_0}{\partial p} = \frac{\alpha m}{2\pi}\int_0^\infty \frac{ds}{s}\int_0^1 du\, e^{-isu^2m^2}\left\{(-is)u(1-u)(-2p)\cdot\right.$$

$$\left.\cdot e^{-isu(1-u)(m^2-p^2)}\left[2 + (1-u)\frac{p}{m}\right] + e^{-isu(1-u)(m^2-p^2)}\frac{1-u}{m}\right\}$$

which gives the expression on the mass shell (without c.t.)

$$\frac{\partial \Sigma_0}{\partial \not p}(\not p=-m) = \frac{\alpha m}{2\pi}\int_0^\infty \frac{ds}{s}\int_0^1 du\, e^{-isu^2m^2}\left\{-2ismu(1-u^2)+\frac{1-u}{m}\right\}$$

In order to eliminate this as well, we choose the second part of the counter terms as

$$c.t.^{(2)} = -(m+\gamma\Pi)\left[\frac{1-u}{m} - 2imu(1-u^2)s\right] \tag{3.43b}$$

whereby we make sure that (3.39) still remains valid, by using the factor $(m+\gamma\Pi)$.

Our end result, then, for the renormalized mass operator in a constant external magnetic field reads:

$$\Sigma(\Pi) = \frac{\alpha m}{2\pi}\int_0^\infty \frac{ds}{s}\int_0^1 du\; e^{-isu^2m^2}\Big\{\Delta^{-\frac{1}{2}}e^{-is\phi}\Big[1+ \tag{3.44}$$

$$e^{-2i\sigma^3 y} + (1-u)e^{-2i\sigma^3 y}\frac{\gamma\Pi}{m} + (1-u)\left(\frac{1-u}{\Delta}+\frac{u}{\Delta}\frac{\sin y}{y}e^{-i\sigma^3 y}-\right.$$

$$\left.-e^{-2i\sigma^3 y}\right)\frac{\gamma\Pi_\perp}{m}\Big] - (1+u)$$

$$-(m+\gamma\Pi)\left[\frac{1-u}{m} - 2imu(1-u^2)s\right]\Big\}$$

One can check that in the free field case the conventional spectral representation of the mass operator is obtained [29,39]:

$$\Sigma_0(p) = -(\not p+m)^2\frac{\alpha}{4\pi}\int_{\to m}^\infty \frac{dM}{M}\left(1-\frac{m^2}{M^2}\right)\cdot$$

$$\cdot\left\{\frac{1-\frac{2mM}{(M-m)^2}}{\not p+M-i\varepsilon} + \frac{1+\frac{2mM}{(M+m)^2}}{\not p-M+i\varepsilon}\right\} \tag{3.45}$$

We observe that for $M \to m$ the well-known infrared divergence appears, which has its origin in the masslessness of the photon.

As an important application of the mass operator with regard to the renormalization of the 2-loop effective Lagrangian, we are now going to explicitly evaluate the mass displacement

$$\delta m = \Sigma^{nr}(\not p = -m)$$

where Σ^{nr} denotes the unrenormalized mass operator, i.e., Σ without counter terms. From equation (3.42) we read off immediately:

$$\delta m = \frac{\alpha m}{2\pi} \int_0^\infty \frac{ds}{s} \int_0^1 du\ (1+u)\, e^{-isu^2(m^2-i\varepsilon)}$$

which apperently diverges for $s \to o$ and therefore cannot be used in this form. Furthermore, for $u = o$, the exponential damping of the integrand vanishes, i.e., for $u = o$, the s-integration would be divergent, at the upper limit as well. To obtain a regularization scheme consistent with section (7) we first reverse the variable substitution (3.12), and introduce finite lower limits for both integrals:

$$\delta m(s_0, s_0') = \frac{\alpha m}{2\pi} \int_{s_0}^\infty ds_1 \int_{s_0'}^\infty ds_2\, \frac{(2s_1+s_2)}{(s_1+s_2)^3}\, e^{-i(m^2-i\varepsilon)\frac{s_1^2}{s_1+s_2}} \tag{3.46}$$

Here, s_1 and s_2 are again the integration variables from the proper time representations of the electron and photon propagator. Now we can put $s_0' = 0$ (but $s_0 > 0$) without making the integral become divergent:

$$\delta m(s_0) = \frac{\alpha m}{2\pi} \int_{s_0}^\infty ds_1 \int_0^\infty ds_2\, \frac{(2s_1+s_2)}{(s_1+s_2)^3}\, e^{-i(m^2-i\varepsilon)\frac{s_1^2}{s_1+s_2}} \tag{3.47}$$

In this form, i.e., as a function of the lower limit of the proper time integration of the electron propagator we shall use δm later on.

To calculate the (convergent) integral (3.47), we introduce a new substitution for the s_2-integration

$$s_2 \longrightarrow u := \frac{s_1}{s_1+s_2}$$

which, with

$$s_2(u) = \frac{s_1}{u} - s_1 \quad , \quad ds_2 = -\frac{s_1}{u^2}\, du$$

leads to

$$\delta m(s_0) = \frac{\alpha m}{2\pi} \int_{s_0}^{\infty} \frac{ds_1}{s_1} \int_0^1 du\ (1+u)\, e^{-i(m^2-i\varepsilon)us_1}$$

The u-integration is now elementary, and, with the aid of

$$\int dx\ e^{-ax} = -\frac{1}{a} e^{-ax}$$

$$\int dx\ x e^{-ax} = -\frac{x}{a} e^{-ax} - \frac{1}{a^2} e^{-ax}$$

we get

$$\delta m(s_0) = \frac{\alpha m}{2\pi} \Big\{ -\frac{2}{im^2} \int_{s_0}^{\infty} ds_1 \frac{e^{-im^2 s_1}}{s_1^2} - \frac{1}{(im^2)^2} \int_{s_0}^{\infty} ds_1 \frac{e^{-im^2 s_1}}{s_1^3}$$

$$+ \frac{1}{im^2} \int_{s_0}^{\infty} \frac{ds_1}{s_1^2} + \frac{1}{(im^2)^2} \int_{s_0}^{\infty} \frac{ds_1}{s_1^3} \Big\}$$

Now, in the first and second term, we do one, respectively two, integrations by parts, and calculate the integrals in the third and fourth term. The result is:

$$\delta m(s_0) = \frac{\alpha m}{2\pi}\Big[-\frac{e^{-im^2s_0}}{2(im^2s_0)^2} + \frac{3}{2}\frac{e^{-im^2s_0}}{(im^2s_0)} + \frac{1}{im^2s_0}$$

$$+\frac{1}{2(im^2s_0)^2} + \frac{3}{2}\int_{s_0}\frac{ds_1}{s_1}e^{-i(m^2-i\varepsilon)s_1}\Big]$$

Since we are interested in the limit $s_o \to o$, we expand the exponentials and disregard all terms which vanish for $s_o \to o$; this gives

$$\delta m(s_0) = \frac{3\alpha m}{4\pi}\Big[\int_{s_0}^{\infty} ds_1\frac{e^{-i(m^2-i\varepsilon)s_1}}{s_1} + \frac{5}{6}\Big] + O(s_0)$$

Performing the exponential integral [36] leads to

$$\int_{s_0}^{\infty} ds_1\frac{e^{-i(m^2-i\varepsilon)s_1}}{s_1} = Ei\left(-i(m^2-i\varepsilon)s_1\right)\Big|_{s_0}^{\infty}$$

$$= -Ei\left(-i(m^2-i\varepsilon)s_0\right)$$

which, for $s_o \to o$, reduces to [36]

$$\int_{s_0}^{\infty} ds_1\frac{e^{-i(m^2-i\varepsilon)s_1}}{s_1} = -\left[C + \ln(im^2s_0)\right] + O(s_0)$$

$$= \ln\frac{1}{im^2\gamma s_0} + O(s_0)$$

Here, $C = \ln\gamma$ is Euler's constant. So, our final result for the mass displacement reads:

$$\delta m(s_0) = \frac{3\alpha m}{4\pi}\Big[\ln\frac{1}{i\gamma m^2 s_0} + \frac{5}{6}\Big] \qquad (3.48a)$$

or, for the change in m^2:

$$\delta m^2(s_0) = 2m\,\delta m(s_0) = \frac{3\alpha m^2}{2\pi}\left[\ln\frac{1}{i\gamma m^2 s_0} + \frac{5}{6}\right] \tag{3.48b}$$

This equation corresponds exactly to Schwinger's [3] and Ritus' [4] results, but was achieved by completely different methods.

(4) The Polarization Tensor in a Constant External Magnetic Field

A further important building block for higher-order processes in QED is the polarization tensor, which we shall calculate to the order α in this section. First, however, we consider the completely dressed photon propagator without external field, defined by

$$D'_{+\mu\nu}(x-x') = i\langle 0|T\,A_\mu(x)\,A_\nu(x')|0\rangle \tag{4.1}$$

and whose Fourier transform can be written as

$$D'_{+\mu\nu}(k) = \int d^4x\, e^{-ik\cdot x}\, D'_{+\mu\nu}(x)$$

The perturbation series of D'_+ with respect to α thus contains the graphs

$$D'_{+\mu\nu}(k) =$$

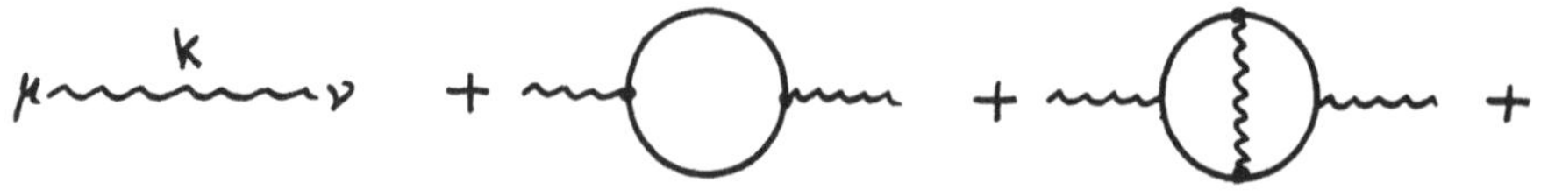

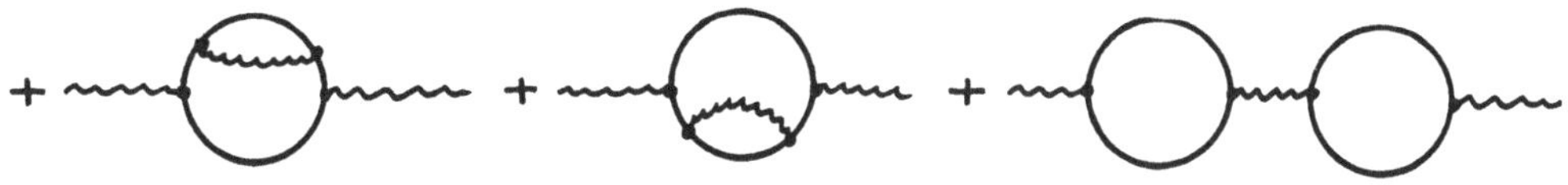

where the last diagram is one-particle-reducible, i.e., one only has to cut one inner line in order to get two simpler diagrams. So, analogous to the definition of the mass operator in the last section, it is convenient here to introduce a proper self-energy part of the photon or polarization tensor $\Pi_{\mu\nu}(k)$ as the sum of all contributing one-particle-irreducible graphs without external propagators. By explicitly calculating the diagram of second order in e given by

$$\Pi^{(2)}_{\mu\nu}(k) = -ie^2 \; tr \int \frac{d^4p}{(2\pi)^4} \gamma_\mu G_+(p) \gamma_\nu G_+(p-k)$$

we shall see that $\Pi^{(2)}_{\mu\nu}(k)$ can be written as

$$\Pi^{(2)}_{\mu\nu}(k) = (g_{\mu\nu} k^2 - k_\mu k_\nu)\, \Pi^{(2)}(k^2)$$

with a scalar polarization function $\Pi^{(2)}(k^2)$ for which we wish to assume a power series expansion of the form

$$\Pi^{(2)}(k^2) = \Pi_1 + \Pi_2 k^2 + \Pi_3 (k^2)^2 + \text{-----}$$

By iteration of $\Pi^{(2)}_{\mu\nu}$, we get for the photon propagator up to order e^2:

$$D'^{(2)}_{+\,\mu\nu}(k) = \frac{g_{\mu\nu}}{k^2(1+\Pi^{(2)}(k^2))} + \text{longitudinal terms}$$

$$= \frac{g_{\mu\nu}}{k^2(1+\Pi_1+\Pi_2 k^2 + \text{---})} + \text{Long.}$$

which, near the mass shell $k^2 \approx 0$ (note that the photon pole has not been shifted by the interaction), reduces to

$$D'^{(2)}_{+\mu\nu}(k^2 \approx 0) = \frac{g_{\mu\nu}}{k^2(1+\Pi^{(2)}(0))} + \text{Long.} = \frac{Z_3\, g_{\mu\nu}}{k^2} + \text{Long.}$$

with

$$Z_3 = (1+\Pi^{(2)}(0))^{-1} = (1+\Pi_1)^{-1}$$

In order for the normalization of the wave function to be retained, we again must require that the residue of the pole at $k^2 = 0$ be one; this can be achieved by considering the field A_μ in (4.1) as a bare field $A_{\mu o}$ which, according to

$$A_\mu = Z_3^{-1/2} A_{\mu o}$$

is related to the renormalized A_μ, so that the photon propagator calculated with it

$$D'^{(2)}_{+\mu\nu\,\text{ren}} = \frac{D'^{(2)}_{+\mu\nu}}{Z_3}$$

$$= \frac{1}{(1+\Pi_1)^{-1}} \cdot \frac{(1+\Pi_1)^{-1} g_{\mu\nu}}{k^2\left[1+(1+\Pi_1)^{-1}(\Pi_2 k^2+\dots)\right]} + \text{Long.}$$

$$= \frac{g_{\mu\nu}}{k^2[1+(\Pi_2 k^2+\dots)]} + O(e^4) + \text{Long.}$$

has the residue one. The associated polarization function is then

$$\Pi^{(2)}_{ren}(k^2) = \Pi_2 k^2 + \Pi_3 (k^2)^2 + \dots$$

$$= \Pi^{(2)}(k^2) - \Pi^{(2)}(k^2=0)$$

Following our line of reasoning in the last section, we can obtain the renormalized polarization tensor, which we now again shall denote be $\Pi^{(2)}_{\mu\nu}$, by replacing the original definition of $\Pi^{(2)}_{\mu\nu}(k)$ by

$$\Pi^{(2)}_{\mu\nu}(k) = -ie^2 \, tr \int \frac{d^4p}{(2\pi)^4} \gamma_\mu G_+(p) \gamma_\nu G_+(p-k) + c.t. \qquad (4.1')$$

and, after evaluating the integral and the trace, choosing the counter terms c.t. in such a way that

$$\Pi^{(2)}_{\mu\nu}(k^2=0) = 0 \qquad (4.2)$$

After these observations, we want now to explicitly calculate $\Pi^{(2)}_{\mu\nu}$ in the presence of a constant magnetic field, with reference to [6]. Thereby we shall limit ourselves to the first non-trivial graph of the perturbation series, i.e., to

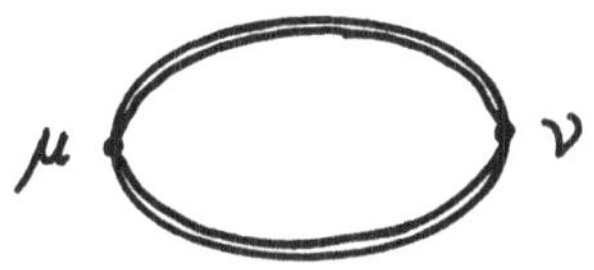

but shall again take into account the external field to all orders of the coupling constant. With the practical notation

$$\langle f(p) \rangle := \int \frac{d^4p}{(2\pi)^4} f(p) \qquad (4.3)$$

we then have to evaluate

$$\Pi^{(2)}_{\mu\nu}(k) = -ie^2 \, tr < \gamma_\mu \, g(p) \, \gamma_\nu \, g(p-k) > + c.t. \qquad (4.4)$$

whereby g is again given by

$$g(p) = i \int_0^\infty ds \, e^{-is[m^2 + p_\parallel^2 + (\tan z / z) p_\perp^2]}$$

$$\cdot \frac{e^{i\sigma^3 z}}{\cos z} \left\{ m - \gamma p_\parallel - \frac{e^{-i\sigma^3 z}}{\cos z} \gamma p_\perp \right\}$$

with z = eBs. Substitution results in

$$\Pi^{(2)}_{\mu\nu}(k) = ie^2 \int_0^\infty ds_1 \int_0^\infty ds_2 \left\langle \exp\{-is_1[m^2 + p_\parallel^2 + (\tan z_1/z_1) p_\perp^2] - \right.$$

$$- is_2 [m^2 + (p-k)_\parallel^2 + (\tan z_2/z_2)(p-k)_\perp^2]\} \cdot \frac{1}{\cos z_1 \cdot \cos z_2} \cdot$$

$$tr \{ \gamma_\mu [(m - \gamma p_\parallel) e^{i\sigma^3 z_1} - \frac{\gamma p_\perp}{\cos z_1}] \gamma_\nu [(m - \gamma (p-k)_\parallel) e^{i\sigma^3 z_2} -$$

$$\left. - \frac{\gamma (p-k)_\perp}{\cos z_2}]\} \right\rangle + c.t. \qquad (4.5)$$

with $z_1 = eBs_1$ and $z_2 = eBs_2$.

Now we introduce new variables of integration

$$s_1 =: s \, \frac{1-v}{2}$$

$$s_2 =: s \, \frac{1+v}{2} \qquad (4.6)$$

and get

$$S = s_1 + s_2$$
$$v = \frac{s_2 - s_1}{s_2 + s_1}$$
$$z_1 = eBs_1 = eBs\,\frac{1-v}{2} =: z\,\frac{1-v}{2} =: \xi \qquad (4.7)$$
$$z_2 = eBs_2 = eBs\,\frac{1+v}{2} =: z\,\frac{1+v}{2} =: \eta$$

as well as

$$\int_0^\infty ds_1 \int_0^\infty ds_2 \;.... = \int_0^\infty s\,ds \int_{-1}^{1} \frac{dv}{2} \;.... \qquad (4.8)$$

To simplify (4.5), we first note that the exponential function appearing in it can be put in the form

$$\exp\left\{-is_1\left[m^2 + p_\parallel^2 + \frac{\tan\xi}{\xi}\,p_\perp^2\right] - is_2\left[m^2 + (p-k)_\parallel^2 + \frac{\tan\eta}{\eta}\,(p-k)_\perp^2\right]\right\}$$
$$= \exp\left\{-is(\varphi_0 + \varphi_1)\right\} \qquad (4.9)$$

with

$$\varphi_0 = m^2 + \frac{1-v^2}{4}\,k_\parallel^2 + \frac{\cos zv - \cos z}{2z\sin z}\,k_\perp^2 \qquad (4.10)$$

and

$$\varphi_1 = \left(p_\parallel - \frac{1+v}{2}k_\parallel\right)^2 + \frac{\tan\xi + \tan\eta}{z}\left(p_\perp - \frac{\tan\eta}{\tan\xi + \tan\eta}k_\perp\right)^2 \qquad (4.11)$$

Now we are allowed to put (4.9) into (4.5) and obtain

$$\Pi_{\mu\nu}(k) = ie^2 \int_0^\infty s\,ds \int_{-1}^{1} \frac{dv}{2}\, e^{-is\varphi_0}\, \frac{1}{\cos\xi\cdot\cos\eta} \qquad (4.12)$$

$$\cdot\, tr < e^{-is\varphi_1}\left\{\gamma_\mu\left[(m-\gamma p_\parallel)e^{i\sigma^3\xi} - \frac{\gamma p_\perp}{\cos\xi}\right]\gamma_\nu\left[(m-\gamma(p-k)_\parallel)e^{i\sigma^3\eta} - \right.\right.$$
$$\left.\left. - \frac{\gamma(p-k)_\perp}{\cos\eta}\right]\right\} > + \text{c.t.}$$

We can now perform the p-integration very easily, by shifting the integration variable and using equation (3.19). The simplest integral needed in (4.12) is

$$\langle e^{-is\varphi_1}\rangle = \int\frac{d^4p}{(2\pi)^4}\exp[-is\{(p_\parallel-\tfrac{1+v}{2}k_\parallel)^2+\frac{\tan\xi+\tan\eta}{z}(p_\perp-\frac{\tan\eta}{\tan\xi+\tan\eta}k_\perp)^2\}]$$
$$= \frac{(-i)}{(4\pi)^2}\,\frac{1}{s^2}\,\frac{z}{\sin z}\cos\xi\cos\eta \tag{4.13}$$

Furthermore

$$\langle e^{-is\varphi_1}p_\parallel\rangle = \int\frac{d^4p}{(2\pi)^4}\exp[-is\{(p_\parallel-\tfrac{1+v}{2}k_\parallel)^2+\frac{\tan\xi+\tan\eta}{z}(p_\perp-\frac{\tan\eta}{\tan\xi+\tan\eta}k_\perp)^2\}]\,p_\parallel$$
$$= \frac{1+v}{2}k_\parallel\langle e^{-is\varphi_1}\rangle \tag{4.14}$$

since the integral vanishes over an odd function. Analogously we get

$$\langle e^{-is\varphi_1}p_\perp\rangle = \frac{\tan\eta}{\tan\xi+\tan\eta}k_\perp\langle e^{-is\varphi_1}\rangle$$

$$\langle e^{-is\varphi_1}p_{\parallel\mu}p_{\parallel\nu}\rangle = \langle e^{-is\varphi_1}\rangle\left[\left(\frac{1+v}{2}\right)^2 k_{\parallel\mu}k_{\parallel\nu}-\frac{i}{2s}g^\parallel_{\mu\nu}\right] \tag{4.15}$$

$$\langle e^{-is\varphi_1}\{p_{\parallel\mu}p_{\perp\nu},p_{\perp\mu}p_{\parallel\nu}\}\rangle = \langle e^{-is\varphi_1}\rangle\frac{1+v}{2}\,\frac{\tan\eta}{\tan\xi+\tan\eta}\{k_{\parallel\mu}k_{\perp\nu},k_{\perp\mu}k_{\parallel\nu}\}$$

$$\langle e^{-is\varphi_1}p_{\perp\mu}p_{\perp\nu}\rangle = \langle e^{-is\varphi_1}\rangle\left[\left(\frac{\tan\eta}{\tan\eta+\tan\xi}\right)^2 k_{\perp\mu}k_{\perp\nu}\right.$$
$$\left.-\frac{i}{2s}\,\frac{z}{\tan\xi+\tan\eta}g^\perp_{\mu\nu}\right]$$

whereby we introduced the definitions

$$(g^\parallel_{\mu\nu}) = (g^{\parallel\mu\nu}) = \begin{pmatrix}-1&&&0\\&0&&\\&&0&\\0&&&1\end{pmatrix}$$
$$(g^\perp_{\mu\nu}) = (g^{\perp\mu\nu}) = \begin{pmatrix}0&&&0\\&1&&\\&&1&\\0&&&0\end{pmatrix} \tag{4.16}$$

With the help of (4.13) we can now eliminate the factor $(\cos\xi\cos\eta)^{-1}$ which leads to

$$\Pi_{\mu\nu}(k) = \frac{\alpha}{2\pi}\int_0^\infty \frac{ds}{s}\int_{-1}^{1}\frac{dv}{2}\, e^{-is\varphi_0}\, I_{\mu\nu}\,\frac{z}{\sin z} + c.t. \tag{4.17}$$

with

$$I_{\mu\nu} \equiv 2\,\frac{1}{4\langle e^{-is\varphi_0}\rangle}\, tr\langle e^{-is\varphi_1}\{\gamma_\mu[(m-\gamma p_\parallel)e^{i\sigma^3\xi} - \frac{\gamma p_\perp}{\cos\xi}]\gamma_\nu \cdot$$

$$\cdot[(m-\gamma(p-k)_\parallel)e^{i\sigma^3\eta} - \frac{\gamma(p-k)_\perp}{\cos\eta}]\}\rangle$$

With this, we must evaluate the following trace

$$tr\langle e^{-is\varphi_1}\gamma_\mu(m e^{i\sigma^3\xi} - \gamma p_\parallel e^{i\sigma^3\xi} - \frac{\gamma p_\perp}{\cos\xi})\gamma_\nu(m e^{i\sigma^3\eta} - \gamma(p-k)_\parallel e^{i\sigma^3\eta} - \frac{\gamma(p-k)_\perp}{\cos\eta})\rangle$$

$$\begin{aligned} = \; & m^2\, tr\langle e^{-is\varphi_1}\gamma_\mu e^{i\sigma^3\xi}\gamma_\nu e^{i\sigma^3\eta}\rangle \\ & + tr\langle e^{-is\varphi_1}\gamma_\mu \gamma p_\parallel e^{i\sigma^3\xi}\gamma_\nu \gamma(p-k)_\parallel e^{i\sigma^3\eta}\rangle \\ & + tr\langle e^{-is\varphi_1}\gamma_\mu \gamma p_\parallel e^{i\sigma^3\xi}\gamma_\nu \gamma(p-k)_\perp\rangle \frac{1}{\cos\eta} \\ & + tr\langle e^{-is\varphi_1}\gamma_\mu \gamma p_\perp \gamma_\nu \gamma(p-k)_\parallel e^{i\sigma^3\eta}\rangle \frac{1}{\cos\xi} \\ & + tr\langle e^{-is\varphi_1}\gamma_\mu \gamma p_\perp \gamma_\nu \gamma(p-k)_\perp\rangle \frac{1}{\cos\xi\cdot\cos\eta} \end{aligned} \tag{4.18}$$

Before we begin to solve this problem, we derive two useful trace relations which will prove helpful. The first reads:

$$\frac{1}{4}tr\{e^{i\sigma^3 z}\gamma_\mu\gamma_\nu\} = -\cos z\; g_{\mu\nu} + \left(\frac{F}{B}\right)_{\mu\nu}\cdot\sin z \tag{4.19}$$

By way of proof, we take advantage of the fact that only the components $F_{12} = -F_{21} = B$ of the field strength tensor do not vanish, i.e., that

$$\left(\frac{F}{B}\right)_{\mu\nu} = g_{1\mu}\, g_{2\nu} - g_{1\nu}\, g_{2\mu}$$

is valid.

With the formulae given in Appendix A we then obtain

$$\frac{1}{4}\,\mathrm{tr}\{e^{i\sigma^3 z}\gamma_\mu\gamma_\nu\} = \frac{1}{4}\,\mathrm{tr}\{(\cos z + i\sigma^3 \sin z)\,\gamma_\mu\gamma_\nu\}$$

$$= -\cos z\; g_{\mu\nu} + \left(\frac{F}{B}\right)_{\mu\nu} \sin z$$

In addition, we need the relation

$$\frac{1}{4}\,\mathrm{tr}\{e^{i\sigma^3 z}\gamma_\mu\gamma_\nu\gamma_\lambda\gamma_\sigma\} = \cos z\;[g_{\mu\nu}\, g_{\lambda\sigma} - g_{\mu\lambda}\, g_{\nu\sigma} + g_{\mu\sigma}\, g_{\nu\lambda}] - \tag{4.20}$$

$$-\frac{1}{B}\sin z\left[F_{\mu\nu}\, g_{\lambda\sigma} - F_{\mu\lambda}\, g_{\nu\sigma} + F_{\mu\sigma}\, g_{\nu\lambda} - F_{\nu\sigma}\, g_{\mu\lambda} + F_{\nu\lambda}\, g_{\mu\sigma} + F_{\lambda\sigma}\, g_{\mu\nu}\right]$$

Supplied with (4.19) and (4.20) we can now begin calculating the individual sums of (4.18), which we shall denote as $\mathrm{tr}\, S_i$, $i = 1...5$. First we get

$$\frac{\mathrm{tr}\, S_1}{4m^2\langle e^{-is\varphi_0}\rangle} = \frac{1}{4}\,\mathrm{tr}\langle e^{-is\varphi_0}\gamma_\mu\, e^{-i\sigma^3\xi}\gamma_\nu\, e^{i\sigma^3\eta}\rangle\langle e^{-is\varphi_0}\rangle^{-1}$$

$$= \cos\eta\;\left[-\cos\xi\; g_{\mu\nu} + \left(\frac{F}{B}\right)_{\nu\mu}\sin\xi\right]$$

$$-\sin\eta\;\underbrace{\frac{1}{4}\,\mathrm{tr}\{e^{i\sigma^3\xi}\gamma_\nu\gamma_1\gamma_2\gamma_\mu\}}_{=:\,A}$$

Here, we used the cyclicity of the trace and the antisymmetry of $\gamma_1\gamma_2$ with respect to the indices 1 and 2. In the second sum we now need

$$A \underset{(4.20)}{=} \cos\xi \; [g_{\nu 1} g_{2\mu} - g_{\nu 2} g_{1\mu} + g_{\nu\mu} g_{12}]$$
$$- \frac{1}{B} \sin\xi [F_{\nu 1} g_{2\mu} - F_{\nu 2} g_{1\mu} + F_{\nu\mu} g_{12} - F_{1\mu} g_{\nu 2} + F_{12} g_{\nu\mu} + F_{2\mu} g_{\nu 1}]$$
$$= -\left(\frac{F}{B}\right)_{\mu\nu} \cos\xi - \sin\xi [-g_{\nu 2} g_{2\mu} - g_{1\nu} g_{1\mu} - g_{\mu 2} g_{\nu 2} + g_{\nu\mu} - g_{\mu 1} g_{\nu 1}]$$
$$= -\left(\frac{F}{B}\right)_{\mu\nu} \cos\xi - \sin\xi \, g_{\mu\nu} + 2 \sin\xi \; g^{\perp}_{\mu\nu}$$

We again used the special form of the field strength tensor in the second line. If this is put into the above equation and the addition theorem of trigometric functions is used, it follows that

$$\frac{tr\, S_1}{4m^2 \langle e^{-is\varphi_1}\rangle} = -\cos z\, g_{\mu\nu} + \left(\frac{F}{B}\right)_{\mu\nu} \sin zv - 2 \sin\xi \sin\eta \; g^{\perp}_{\mu\nu}$$

It should be noted that the second term on the right-hand side is an odd function in v, which after substitution into (4.17) makes no contribution. If we still use

$$g_{\mu\nu} = g^{\parallel}_{\mu\nu} + g^{\perp}_{\mu\nu} \tag{4.21}$$

then, again thanks to the addition theorem

$$\frac{tr\, S_1}{4 \langle e^{-is\varphi_1}\rangle} = -m^2 \left[\cos zv\, g^{\perp}_{\mu\nu} + \cos z\, g^{\parallel}_{\mu\nu} + \text{odd function}\right] \tag{4.22}$$

To further evaluate (4.18), we must go back to the integrals (4.14) and (4.15); for tr S_2, we get, for example

$$\frac{tr\, S_2}{4 \langle e^{-is\varphi_1}\rangle} = \frac{1}{4} \langle e^{-is\varphi_1}\rangle^{-1} tr \langle e^{-is\varphi_1} \gamma_\mu \gamma p_\parallel e^{i\sigma^3 \xi} \gamma_\nu \gamma(p-k)_\parallel e^{i\sigma^3 \eta} \rangle$$
$$=: C^{\alpha\beta} D_{\alpha\beta}$$

with

$$C^{\alpha\beta} = \langle e^{-is\varphi_1}\rangle^{-1}\left\{\langle e^{-is\varphi_1} p_{\parallel}^{\alpha} p_{\parallel}^{\beta}\rangle - k_{\parallel}^{\beta}\langle e^{-is\varphi_1} p_{\parallel}^{\alpha}\rangle\right\}$$

$$= -\frac{1-v^2}{4} k_{\parallel}^{\alpha} k_{\parallel}^{\beta} - \frac{i}{s} g_{\parallel}^{\alpha\beta}$$

and

$$D^{\alpha\beta} = \frac{1}{4}\, tr\left\{\gamma_{\mu}\gamma_{\parallel\alpha} e^{i\sigma^3\xi}\gamma_{\nu}\gamma_{\parallel\beta} e^{i\sigma^3\eta}\right\}.$$

Due to the cyclicity of the trace it then follows

$$\frac{tr\, S_2}{4\langle e^{-is\varphi_1}\rangle} = \frac{1}{4}\, tr\left\{\left(-\frac{1-v^2}{4}\right)\gamma k_{\parallel} e^{i\sigma^3\xi}\gamma_{\nu}\gamma k_{\parallel} e^{i\sigma^3\eta}\gamma_{\mu} - \right.$$

$$\left. -\frac{i}{2s}\gamma_{\parallel}^{\lambda} e^{i\sigma^3\xi}\gamma_{\nu}\gamma_{\parallel\lambda} e^{i\sigma^3\eta}\gamma_{\mu}\right\}.$$

It can easily be shown that

$$\gamma k_{\parallel} e^{i\sigma^3\xi} = e^{i\sigma^3\xi}\gamma k_{\parallel}$$

$$\gamma k_{\parallel}\gamma_{\nu} = -2k_{\parallel\nu} - \gamma_{\nu}\gamma k_{\parallel}$$

$$\gamma_{\parallel}^{\lambda} e^{i\sigma^3\xi} = e^{i\sigma^3\xi}\gamma_{\parallel}^{\lambda}$$

and
$$\gamma_{\parallel}^{\lambda}\gamma_{\nu} = -\gamma_{\nu}\gamma_{\parallel}^{\lambda} - 2g_{\parallel\ \nu}^{\ \lambda}$$

is valid, and, substituted into the above equation, leads to

$$\frac{tr\, S_2}{4\langle e^{-is\varphi_1}\rangle} = \left(-\frac{1-v^2}{4}\right)\frac{1}{4}\, tr\left\{e^{i\sigma^3\xi}(-\gamma_{\nu}\gamma k_{\parallel} - 2k_{\parallel\nu})\gamma k_{\parallel} e^{i\sigma^3\eta}\gamma_{\mu}\right\}$$

$$-\frac{i}{2s}\frac{1}{4}\, tr\left\{e^{i\sigma^3\xi}(-\gamma_{\nu}\gamma_{\parallel}^{\lambda} - 2g_{\parallel\ \nu}^{\ \lambda})\gamma_{\parallel\lambda} e^{i\sigma^3\eta}\gamma_{\mu}\right\}$$

With
$$\gamma k_{\parallel}\gamma k_{\parallel} = -k_{\parallel}^2$$

and
$$\gamma_{\parallel}^{\lambda}\gamma_{\parallel\lambda} = -2$$

this becomes

$$\frac{tr\, S_2}{4\langle e^{-is\varphi_1}\rangle} = \left(-\frac{1-v^2}{4}\right)\frac{1}{4}\, tr\left\{e^{i\sigma^3\xi}(\gamma_\nu k_\parallel - 2k_{\parallel\nu}\gamma k_\parallel)e^{i\sigma^3\eta}\gamma_\mu\right\}$$

$$-\frac{i}{2s}\frac{1}{4}\, tr\left\{e^{i\sigma^3\xi}\, 2\gamma_{\perp\nu}\, e^{i\sigma^3\eta}\gamma_\mu\right\} \quad (4.23)$$

$$=: \mathcal{B}_1 + \mathcal{B}_2 + \mathcal{B}_3$$

where, with the help of (4.22) we can put the B's into the form

$$\mathcal{B}_1 \equiv \left(-\frac{1-v^2}{4}\right)K_\parallel^2\, \frac{1}{4}\, tr\left\{e^{i\sigma^3\xi}\gamma_\nu e^{i\sigma^3\eta}\gamma_\mu\right\}$$

$$= \frac{1-v^2}{4}K_\parallel^2\left[\cos zv\, g^\perp_{\mu\nu} + \cos z\, g^\parallel_{\mu\nu}\right] + \text{odd function}$$

$$\mathcal{B}_2 \equiv -2\left(-\frac{1-v^2}{4}\right)K_{\parallel\nu}K_\parallel^\alpha\, \frac{1}{4}\, tr\left\{e^{i\sigma^3\xi}\gamma_\alpha e^{i\sigma^3\eta}\gamma_\mu\right\} \quad (4.24)$$

$$= 2\left(-\frac{1-v^2}{4}\right)K_{\parallel\mu}K_{\parallel\nu}\cos z + \text{odd function}$$

$$\mathcal{B}_3 \equiv -\frac{i}{s}\frac{1}{4}\, tr\left\{e^{i\sigma^3\xi}\gamma_{\perp\nu}e^{i\sigma^3\eta}\gamma_\mu\right\}$$

$$= \frac{i}{s}\cos zv\, g^\perp_{\mu\nu} + \text{odd function}$$

Evaluation of the remaining terms of (4.18) is accomplished analogously and following a longer but elementary calculation, leads to the following expression for the function $I_{\mu\nu}$:

$$I_{\mu\nu} = 2\sum_{i=1}^{5}\frac{tr\, S_i}{4\langle e^{-is\varphi_1}\rangle} \quad (4.25)$$

$$= \left(-2m^2 + \frac{1-v^2}{2}K_\parallel^2\right)(\cos z\, g^\parallel_{\mu\nu} + \cos zv\, g^\perp_{\mu\nu}) - (1-v^2)\cos z\, K_{\parallel\mu}K_{\parallel\nu} +$$

$$+ \frac{2i}{s}\left(\cos zv\, g^\perp_{\mu\nu} + \frac{z}{\sin z}\, g^\parallel_{\mu\nu}\right)$$

$$- (\cos zv - v\cot z\cdot\sin zv)\left[K_\mu K_\nu - K_{\perp\mu}K_{\perp\nu} - K_{\parallel\mu}K_{\parallel\nu}\right]$$

$$+ \frac{\cos zv - \cos z}{\sin^2 z}\left[g_{\mu\nu}K_\perp^2 - 2K_{\perp\mu}K_{\perp\nu}\right]$$

Further simplification of (4.17) can be achieved by integration by parts

$$\int_0^\infty ds \frac{i}{s^2} e^{-is\varphi_0} \frac{z}{\sin z} (\cos z\, g^{\parallel}_{\mu\nu} + \cos zv\, g^{\perp}_{\mu\nu}) = BT +$$

$$+ i \int_0^\infty \frac{ds}{s} \Big[\frac{z}{\sin z} (\cos z\, g^{\parallel}_{\mu\nu} + \cos zv\, g^{\perp}_{\mu\nu}) \frac{d}{ds} (e^{-is\varphi_0})$$

$$+ e^{-is\varphi_0} \frac{z}{\sin z} \frac{d}{ds} (\cos z\, g^{\parallel}_{\mu\nu} + \cos zv\, g^{\perp}_{\mu\nu})$$

$$+ e^{-is\varphi_0} (\cos z\, g^{\parallel}_{\mu\nu} + \cos zv\, g^{\perp}_{\mu\nu}) \frac{d}{ds} \Big(\frac{z}{\sin z} \Big) \Big]$$

$$=: BT + i \int_0^\infty \frac{ds}{s} [F_1 + F_2 + F_3]$$

(BT = boundary terms). Differentiation gives

$$F_1 = \frac{z}{\sin z} (\cos z\, g^{\parallel}_{\mu\nu} + \cos zv\, g^{\perp}_{\mu\nu})(-i)\Big[m^2 + \frac{1-v^2}{4} k_{\parallel}^2 -$$

$$- \Big(\frac{v \sin zv + \cot z \cdot \cos zv}{2 \sin z} - \frac{1}{2 \sin^2 z} \Big) k_{\perp}^2 \Big]$$

$$F_2 = -\frac{1}{s} e^{-is\varphi_0} \frac{z}{\sin z} [z \sin z\, g^{\parallel}_{\mu\nu} + vz \sin vz\, g^{\perp}_{\mu\nu}]$$

$$F_3 = e^{-is\varphi_0} (\cos z\, g^{\parallel}_{\mu\nu} + \cos zv\, g^{\perp}_{\mu\nu}) \frac{1}{s} \frac{z}{\sin z} \Big[1 - z \frac{\cos z}{\sin z} \Big]$$

so that, after a short intermediary calculation, we finally get

$$\int_0^\infty \frac{ds}{s} \frac{z}{\sin z} e^{-is\varphi_0} \frac{i}{s} (\cos z\, g^{\parallel}_{\mu\nu} + \cos zv\, g^{\perp}_{\mu\nu}) = BT + \tag{4.26}$$

$$+ \int_0^\infty \frac{ds}{s} \frac{z}{\sin z} e^{-is\varphi_0} \Big\{ \Big[m^2 + \frac{1-v^2}{4} k_{\parallel}^2 - \Big(\frac{v \sin zv + \cot z \cos zv}{2 \sin z} -$$

$$- \frac{1}{2 \sin^2 z} \Big) k_{\perp}^2 \Big] (\cos z\, g^{\parallel}_{\mu\nu} + \cos zv\, g^{\perp}_{\mu\nu}) + \frac{i}{s} \Big[\Big(\cos z - \frac{z}{\sin z} \Big) g^{\parallel}_{\mu\nu} +$$

$$+ \left[(1 - z\cot z)\cos zv - zv \sin zv \right] g^{\perp}_{\mu\nu} \Big] \Big\}$$

Accordingly,

$$\int_{-1}^{1} \frac{dv}{2} e^{-is\varphi_0} \frac{i}{s} \left[(1 - z\cot z)\cos zv - zv\sin zv \right] = BT +$$

$$+ \int_{-1}^{1} \frac{dv}{2} e^{-is\varphi_0} \left[\tfrac{1}{2}(v\cos zv - \cot z \sin zv)\left(v k_{\parallel}^2 + \frac{\sin zv}{\sin z} k_{\perp}^2 \right) \right]. \tag{4.27}$$

With the help of these two integrations by parts, the integral representation (4.17) of the polarization tensor can be put into a very simple form after some rearrangement:

$$\Pi_{\mu\nu}(K) = \frac{\alpha}{2\pi} \int_0^{\infty} \frac{ds}{s} \int_{-1}^{1} \frac{dv}{2} \left\{ e^{-is\varphi_0} \tilde{I}_{\mu\nu} + c.t. \right\} \tag{4.28a}$$

with

$$\tilde{I}_{\mu\nu} = (g_{\mu\nu} K^2 - K_\mu K_\nu) N_0 - (g^{\parallel}_{\mu\nu} K_{\parallel}^2 - K_{\parallel\mu} K_{\parallel\nu}) N_1 + (g^{\perp}_{\mu\nu} K_{\perp}^2 - K_{\perp\mu} K_{\perp\nu}) N_2 \tag{4.28b}$$

and

$$N_0 = \frac{z}{\sin z}\left(\cos zv - v\cot z \sin zv \right) \tag{4.28c}$$

$$N_1 = -z\cot z\left(1 - v^2 + \frac{v\sin zv}{\sin z} \right) + z\frac{\cos zv}{\sin z}$$

$$N_2 = -\frac{z\cos zv}{\sin z} + \frac{zv\cot z \cdot \sin zv}{\sin z} + \frac{2z(\cos zv - \cos z)}{\sin^3 z}$$

Here, we have incorporated the boundary terms BT in the c.t. Now we must set the counter terms so that the renormalization condition for the polarization tensor

$$\lim_{k^2\to 0}\lim_{B\to 0}\Pi_{\mu\nu}(k) \equiv \lim_{k^2\to 0}\lim_{z\to 0}\Pi_{\mu\nu}(k) = 0 \qquad (4.29)$$

is fulfilled. To that end, we first investigate

$$\varphi_0 = m^2 + \frac{1-v^2}{4}k_\parallel^2 + \frac{\cos zv - \cos z}{2z\sin z}k_\perp^2$$

in the case of a vanishing external field; by a power series expansion z = eBs we find

$$\lim_{z\to 0}\frac{\cos zv - \cos z}{2z\sin z} = \frac{1}{4}(1-v^2)$$

so that $k_\perp^2$ can again be joined with $k_\parallel^2$ to make k^2. This was to be expected, since after turning off the magnetic field, a preferred space direction no longer exists. So

$$\lim_{z\to 0}\varphi_0 = m^2 + \frac{1}{4}(1-v^2)k^2 \qquad (4.30)$$

and

$$\lim_{k^2\to 0}\lim_{B\to 0}\varphi_0 = m^2 \qquad (4.31)$$

It follows that

$$\lim_{z\to 0} N_0 = 1-v^2$$

$$\lim_{z\to 0} N_1 = 0$$

$$\lim_{z\to 0} N_2 = 0$$

The vanishing of N_1 and N_2 again garantees that no preferred direction exists for B = 0. So if one chooses the counter terms in (4.28a) in the form

$$c.t. = -e^{-ism^2}(1-v^2)(k^2 g_{\mu\nu} - k_\mu k_\nu)$$

then one has fulfilled the renormalization conditional (4.29). Our end result then reads

$$\Pi_{\mu\nu}(k) = \frac{\alpha}{2\pi}\int_0^\infty \frac{ds}{s}\int_{-1}^{1}\frac{dv}{2}\Big\{e^{-is\varphi_0}\big[(g_{\mu\nu}k^2 - k_\mu k_\nu)N_0 - (g^{\parallel}_{\mu\nu}k_\parallel^2 - k_{\parallel\mu}k_{\parallel\nu})N_1 + (g^{\perp}_{\mu\nu}k_\perp^2 - k_{\perp\mu}k_{\perp\nu})N_2\big] - e^{-ism^2}(1-v^2)(k^2 g_{\mu\nu} - k_\mu k_\nu)\Big\} \tag{4.32}$$

where the N_i are given by (4.28c).

We shall use this representation of the polarization tensor in the 7th section in order to derive a simple expression for the 2-loop effective Lagrangian. But first, we should like to return to the case B = 0 and construct a spectral representation for the polarization tensor.

From (4.32), one gets for a vanishing magnetic field

$$\Pi_{\mu\nu}(k) = (g_{\mu\nu}k^2 - k_\mu k_\nu)\,\Pi(k^2)$$

with

$$\Pi(k^2) = \frac{\alpha}{2\pi}\int_0^\infty \frac{ds}{s}\int_0^1 dv\,(1-v^2)\,e^{-ism^2}\Big[e^{-is\frac{1-v^2}{4}k^2} - 1\Big]$$

where we took advantage of the fact that the integrand in v is even. Next, we perform a partial integration with respect to v

$$\Pi(k^2) = \frac{\alpha}{2\pi}\int_0^\infty \frac{ds}{s}\Big\{\Big[\Big(v - \frac{v^3}{3}\Big)\Big(\exp\big[-is\big(m^2 + \tfrac{1-v^2}{4}k^2\big)\big] - \exp\big[-ism^2\big]\Big)\Big]_{v=0}^{v=1}$$

$$- \int_0^1 dv\, v\left(1-\frac{v^2}{3}\right)(-is)\left(-\tfrac{1}{2}v\right) K^2 \exp\left[-is\left(m^2+\frac{1-v^2}{4}K^2\right)\right] \Big\}$$

$$= -\frac{\alpha K^2}{4\pi} \int_0^1 dv\, v^2\left(1-\frac{v^2}{3}\right)\, i \int_0^\infty ds\, \exp\left[-is\left(m^2+\frac{1-v^2}{4}K^2\right)\right]$$

For the s-integration, $m^2 - i\varepsilon$ again replaces m^2

$$\Pi(K^2) = -\frac{\alpha}{4\pi} K^2 \int_0^1 dv\, v^2\left(1-\tfrac{1}{3}v^2\right)\left[m^2+\frac{1-v^2}{4}K^2 - i\varepsilon\right]^{-1}$$

$$= -\frac{\alpha}{3\pi} K^2 \int_0^1 dv\, \frac{v^2(3-v^2)}{1-v^2}\left[K^2+\frac{4m^2}{1-v^2} - i\varepsilon\right]^{-1} \tag{4.33}$$

Now we use

$$M^2 := \frac{4m^2}{1-v^2}$$

as the new variable of integration and thus get from (4.33) the desired spectral representation of the polarization function:

$$\Pi(k^2) = -\frac{\alpha}{3\pi} k^2 \int_{4m^2}^\infty \frac{dM^2}{M^2}\left(1-\frac{4m^2}{M^2}\right)^{\frac{1}{2}}\left(1+\frac{2m^2}{M^2}\right)\frac{1}{K^2+M^2-i\varepsilon} \tag{4.34}$$

Because

$$\lim_{\varepsilon\to 0+} \frac{1}{X-X_0 \pm i\varepsilon} = \mathcal{P}\frac{1}{x-x_0} \mp i\pi\,\delta(x-x_0)$$

it follows for the imaginary part

$$\operatorname{Im}\,\Pi(K^2) = \frac{\alpha}{3}\left(1+\frac{4m^2}{K^2}\right)^{\frac{1}{2}}\left(1-\frac{2m^2}{K^2}\right). \tag{4.35}$$

In this section we have assembled all the building blocks needed to calculate the one and two - loop effective Lagrangians of spinor QED following in the next chapters.

(5) One-Loop Effective Lagrangian

Our first goal in this section is the derivation of an integral expression for the effective Lagrangian in the one-loop approximation. The central subject we shall be interested in is the vacuum amplitude in the presence of an external field which, in the framework of this approximation can be written as (see section (1))

$$\langle 0_+ | 0_- \rangle_A = e^{iW^{(1)}[A]} = \exp\left[i \int d^4x\, \mathcal{L}^{(1)}(x)\right] \tag{5.1}$$

with

$$\begin{aligned} iW^{(1)}[A] &= -\,\mathrm{Tr}\,\ln\,(1 - e\not{A}\,G_+)^{-1} \\ &= -\,\mathrm{Tr}\,\ln\,(G_+[A]\,/\,G_+[0]) \end{aligned} \tag{5.2}$$

Here $G_+ \equiv G_+[0]$ is the electron propagator in the field-free case, connected with $G_+[A]$ by

$$G_+[A] = G_+(1 - e\not{A}\,G_+)^{-1} \tag{5.3}$$

furthermore, Tr indicates the trace both in spinor and configuration space.

The 1-loop effective action $W^{(1)}$, i.e., the effective Lagrangian $L^{(1)}$, introduced in (5.1) is the formal expression for the effect which an arbitrary number of 'external photon lines' can have on a single Fermion loop, i.e., $W^{(1)}$ and $L^{(1)}$ represent the sum of all graphs of the type (compare with Appendix G)

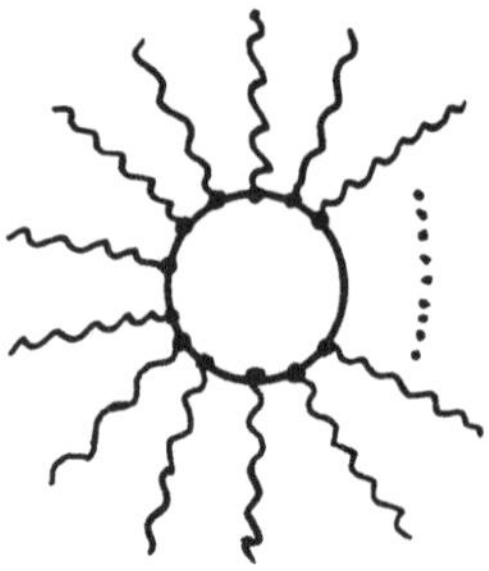

(5.4)

which one abbreviates as

(5.5)

The wavy lines in (5.4) symbolize interactions with the external field which are considered to all orders of the coupling constants and not real photons of the radiation field.

Note that (5.2) vanishes for $F_{\mu\nu} = 0$ so that the vacuum amplitude then becomes

$$\langle 0_+ | 0_- \rangle^{F_{\mu\nu}=0} = e^{iW[F_{\mu\nu}=0]} = 1$$ as it must be.

On the other hand, if $F_{\mu\nu} \neq 0$, then the possiblity exists for $|\langle 0_+|0_-\rangle_A|^2 \neq 1$ to produce electron-positron pairs by means of the external field. As already discussed in the introduction, the probability for this is

$$\begin{aligned} P &= 1 - |\langle 0_+ | 0_- \rangle_A|^2 \\ &= 1 - |e^{iW[A]}|^2 \\ &= 1 - e^{-2\,\mathrm{Im}\,W[A]} \end{aligned}$$

For small values of W, one can expand the exponential function and only consider the linear term:

$$\begin{aligned} P &\approx 1-(1-2\,\mathrm{Im}\,W) \\ &= 2\,\mathrm{Im}\,W \\ &= \int d^4x\,\{2\,\mathrm{Im}\,\mathcal{L}(x)\} \end{aligned}$$

So, for the pair production probability w per unit space and time, it follows:

$$w = 2\,\mathrm{Im}\,\mathcal{L} \tag{5.6}$$

Here, we have always written $W(L)$ rather than $W^{(1)}(L^{(1)})$, to signify that, in order to determine L exactly diagrams with an arbitrary number 1 of loops had to be summed up, i.e., it is

$$\mathcal{L} = \mathcal{L}^{(0)} + \mathcal{L}^{(1)} + \mathcal{L}^{(2)} + \cdots + \mathcal{L}^{(l)} + \cdots \tag{5.7}$$

with the classical Maxwell-Lagrangian $L^{(0)}$.

In section 8, we shall explicitly perform such a summation in the limiting case of strong fields.

But let us now return to $L^{(1)}$ again and look for an explicit representation for this most simple, non-trivial contribution to (5.7)! We note first of all the functional derivative calculated in Appendix D of the $W^{(1)}$ with respect to the potential A_μ:

$$i\,\frac{\delta W^{(1)}[A]}{\delta A_\mu(x)} = -e\,\mathrm{tr}\left[\gamma^\mu G_+(x,x|A)\right] \tag{5.8}$$

In addition we need the proper time representation [3] of the electron propagator:

$$G_+[A] \underset{(2.3)}{=} \frac{\not{\pi} - m}{\not{\pi}^2 - m^2}$$

$$= (m - \not{\pi})\, i \int_0^\infty ds\, e^{-is[m^2 - \not{\pi}^2]} \qquad (5.9)$$

We now show that the Ansatz

$$iW^{(1)} = i\int d^4x\, \mathcal{L}^{(1)} = -\frac{1}{2}\int_0^\infty \frac{ds}{s}\, e^{-ism^2}\, Tr\left[e^{-is\not{\pi}^2}\right] \qquad (5.10)$$

fulfills eq. (5.8) and thereby gives $W^{(1)}$ (to within a constant). We shall later determine this constant so that the action vanishes for vanishing external field as well.

To calculate the derivative from (5.10), we use the chain rule in the form

$$\frac{\delta}{\delta A_\alpha(x)} = \int d^4z\, \frac{\delta \pi_\beta(z)}{\delta A_\alpha(x)} \frac{\delta}{\delta \pi_\beta(z)}$$

$$= -e \frac{\delta}{\delta \pi_\alpha(x)}$$

where $\pi_\beta = p_\beta - eA_\beta$ was used , and we get:

$$i\frac{\delta W^{(1)}[A]}{\delta A_\alpha(x)} = -\frac{1}{2}\int_0^\infty \frac{ds}{s}\, e^{-im^2 s}\, tr \int d^4y\, \frac{\delta}{\delta A_\alpha(x)}\, e^{is\gamma_\mu\gamma_\nu \pi^\mu(y)\pi^\nu(y)}$$

$$= \frac{e}{2}\int_0^\infty \frac{ds}{s}\, e^{-im^2 s}\, tr\int d^4y\; is\gamma^\mu\gamma^\nu \cdot$$

$$\cdot \left\{\frac{\delta}{\delta \pi_\alpha(x)}\, \pi^\mu(y)\pi^\nu(y)\right\} e^{is\gamma_\mu\gamma_\nu \pi^\mu(y)\pi^\nu(y)}$$

With

$$\frac{\delta}{\delta \pi^\alpha(x)}\, \pi^\mu(y)\pi^\nu(y) = \delta^\mu_\alpha\, \delta(x-y)\pi^\nu(y) + \delta^\nu_\alpha\, \delta(x-y)\pi^\mu(y)$$

follows

$$i\,\frac{\delta W^{(1)}[A]}{\delta A_\alpha(x)} = ie\int_0^\infty ds\, e^{-im^2 s}\, tr\left[\gamma^\alpha \langle x|\,\not{\Pi}\, e^{is\not{\Pi}^2}\,|x\rangle\right]$$

$$= -e\; tr\left[\gamma^\alpha \langle x|(m-\not{\Pi})\; i\int_0^\infty ds\, e^{-is(m^2-\not{\Pi}^2)}\,|x\rangle\right]$$

$$\underset{(5.9)}{=} -e\; tr\left[\gamma^\alpha G_+(x,x|A)\right] \qquad (5.11)$$

Here, in the second line we took advantage of the fact that the trace of an odd number of gamma matrices vanishes. With this, the validity of (5.10) is demonstrated, and one can therefore write for the unrenormalized Lagrangian

$$\mathcal{L}^{(1)} = \frac{i}{2}\; tr\int_0^\infty \frac{ds}{s}\, e^{-ism^2}\langle x|\, e^{is\not{\Pi}^2}\,|x\rangle \qquad (5.12)$$

whereby the trace tr only refers to the spinor index. Before further evaluating this expression, we calculate the derivative

$$i\,\frac{\partial\mathcal{L}^{(1)}}{\partial m} = -\frac{1}{2}\int_0^\infty \frac{ds}{s}\,(-2ims)\, e^{-im^2 s}\langle x|\, e^{is\not{\Pi}^2}\,|x\rangle$$

$$= tr\,\langle x|(m-\not{\Pi})\; i\int_0^\infty ds\, e^{-im^2 s}\, e^{is\not{\Pi}^2}\,|x\rangle$$

Here, we again used the fact, that the trace of an odd number of gamma matrices vanishes. A comparison with (5.9) leads now to the simple result

$$i\,\frac{\partial\mathcal{L}^{(1)}}{\partial m} = tr\; G_+(x,x|A) \qquad (5.13)$$

Now we return to eq. (5.12) which, because of

$$\not{\Pi}^2 = -\Pi^2 + \frac{e}{2}\sigma_{\mu\nu}F^{\mu\nu} = -\Pi^2 + eB\sigma^3$$

we can also write as

$$\mathcal{L}^{(1)} = \frac{i}{2}\,\mathrm{tr}\int_0^\infty \frac{ds}{s}\, e^{-ism^2}\, e^{ieBs\sigma^3}\langle x|e^{-is\pi^2}|x\rangle \tag{5.14}$$

In section (2), we found

$$\langle x'|e^{-is\pi^2}|x''\rangle = \phi(x',x'')\int\frac{d^4k}{(2\pi)^4}\, e^{ik(x'-x'')}\,\frac{1}{\cos z}\cdot$$

$$\cdot\exp\left[-is\left(k_\parallel^2 + k_\perp^2 \tan z/z\right)\right]$$

so that, due to $\phi(x,x)=1$, we get the gauge invariant diagonal element

$$\langle x|e^{-is\pi^2}|x\rangle = \frac{1}{\cos z}\int\frac{d^4k}{(2\pi)^4}\exp\left[(-is)\left(k_\parallel^2 + k_\perp^2\tan z/z\right)\right]$$

$$= \frac{1}{\cos z}\,\frac{(-i)}{16\pi^4}\,\frac{\pi^2}{s^2}\,\frac{z}{\tan z} \tag{5.15}$$

Here, the k-integration was performed with the help of (3.19). Substitution into (5.14) gives

$$\mathcal{L}^{(1)} = \frac{1}{8\pi^2}\int_0^\infty\frac{ds}{s^3}\, e^{-ism^2}\, z\cot z\,\frac{\mathrm{tr}\, e^{i\sigma^3 z}}{4\cos z}$$

or, with

$$\mathrm{tr}\, e^{i\sigma^3 z} = \mathrm{tr}\,(\cos z + i\sigma^3\sin z) = 4\cos z$$

follows

$$\mathcal{L}^{(1)} = \frac{1}{8\pi^2}\int_0^\infty\frac{ds}{s^3}\, e^{-ism^2}\,(eBs)\cot(eBs) \tag{5.16}$$

This expression apparently diverges for $s \to 0$ and thus requires renormalization. Before we begin with that, however, we want to give another method for the derivation of (5.16). We begin at

$$G_+(x',x''|A) = \phi(x',x'')\int\frac{d^4k}{(2\pi)^4}\, e^{ik(x'-x'')}\, g(k)$$

with g(k) according to (2.47b) and we calculate the diagonal element (which, again, is gauge invariant)

$$G_+(x,x|A) = \int \frac{d^4k}{(2\pi)^4}\, g(k)$$

$$= i\int_0^\infty ds \int \frac{d^4k}{(2\pi)^4} \exp[-is(m^2+k_{\parallel}^2+k_{\perp}^2 \tan z/z)] \cdot$$

$$\cdot \frac{e^{i\sigma^3 z}}{\cos z}\left[m - \gamma k_{\parallel} - \frac{e^{-i\sigma^3 z}}{\cos z}\gamma k_{\perp}\right]$$

$$\underset{(3.14)}{=} i\int_0^\infty ds\, e^{-ism^2}\left(\frac{-i}{s^2}\frac{1}{16\pi^2}\frac{z}{\tan z}\right)\frac{e^{i\sigma^3 z}}{\cos z}\, m$$

Here, in the second line, the second and third term do not make a contribution, since the integrand in these terms is odd in k. From (5.13) it follows

$$i\frac{\partial \mathcal{L}^{(1)}}{\partial m} = \operatorname{tr} G_+(x,x|A)$$

$$= \frac{1}{4\pi^2}\int_0^\infty \frac{ds}{s^2}\, m e^{-ism^2}\, z\cot z$$

Integration then gives

$$\mathcal{L}^{(1)} = \frac{i}{4\pi^2}\int_0^\infty \frac{ds}{s^2}\int_m^\infty dm'\, m' e^{-is(m'^2-i\varepsilon)}\, z\cot z$$

$$= \frac{i}{4\pi^2}\int_0^\infty \frac{ds}{s^2}\left[\frac{e^{-is(m'^2-i\varepsilon)}}{-2is}\right]_m^\infty z\cot z$$

$$= \frac{1}{8\pi^2}\int_0^\infty \frac{ds}{s^3}\, e^{-ism^2}\, z\cot z$$

i.e., exactly (5.16). We have chosen the integration boundaries so that $L^{(1)}$ vanishes for $m \to \infty$. This is in accord with the physical requirement that for infinite fermion mass all non-linear effects must vanish, since in this case, the creation of virtual electron-positron pairs becomes impossible.

Now we return to (5.16) and attempt to get a finite result inspite of the apparent divergence. First we note that $L^{(1)}$ according to (5.16) for B → o does not, as required, vanish, instead the value

$$\mathcal{L}^{(1)}(B=0) = \frac{1}{8\pi^2}\int_0^\infty \frac{ds}{s^3}\, e^{-im^2 s}$$

arises. Since the addition of a constant of L has no physical meaning, we can subtract this term from (5.16) to get

$$\mathcal{L}^{(1)}(B) = \frac{1}{8\pi^2}\int_0^\infty \frac{ds}{s^3}\, e^{-im^2 s}\left[(eBs)\cot(eBs)-1\right] \tag{5.17}$$

which now does fulfill

$$\mathcal{L}^{(1)}(B=0) = 0$$

but still diverges for s → o. This becomes clear when one expands the above expression:

$$\frac{1}{s^3}\left[(eBs)\cot(eBs)-1\right] = \frac{1}{s^3}\left[1-\frac{1}{3}(eBs)^2-\frac{1}{45}(eBs)^4-\dots-1\right]$$

$$\sim \frac{1}{s} \quad \text{for } s \to o$$

The divergence which stems then from the quadratic term in B can now be gotton rid of by renormalization of charge and field strength. We should keep in mind that in our entire calculations up until now, we should have actually written m_o, e_o, B_o instead of m, e, B, in order to indicate that the bare parameters which first parametrize the theory do not have to coincide with the observable quantities. Furthermore, we recall that the entire Lagrangian in one-loop approximation contains, besides $L^{(1)}$, the classical part

$$\mathcal{L}^{(0)} = -\frac{1}{4} F_{\mu\nu} F^{\mu\nu} = -\frac{1}{2} B^2 \tag{5.18}$$

as well. So we can write

$$\begin{aligned}
\mathcal{L} &= \mathcal{L}^{(0)} + \mathcal{L}^{(1)} \\
&= -\frac{1}{2} B_0^2 + \frac{1}{8\pi^2} \int_0^\infty \frac{ds}{s^3} e^{-im^2 s} \left[(e_0 B_0 s) \cot (e_0 B_0 s) - 1 \right] \\
&= -\frac{1}{2} B_0^2 - \frac{1}{8\pi^2} \int_{s_0}^\infty \frac{ds}{s^3} e^{-im^2 s} \frac{1}{3} (e_0 B_0 s)^2 \\
&\quad + \frac{1}{8\pi^2} \int_0^\infty \frac{ds}{s^3} e^{-im^2 s} \left[(e_0 B_0 s) \cot (e_0 B_0 s) + \frac{1}{3} (e_0 B_0 s) - 1 \right] \\
&= -\frac{1}{2} B_0^2 Z_3^{-1} \\
&\quad + \frac{1}{8\pi^2} \int_0^\infty \frac{ds}{s^3} e^{-im^2 s} \left[(e_0 B_0 s) \cot (e_0 B_0 s) + \frac{1}{3} (e_0 B_0 s)^2 - 1 \right]
\end{aligned} \tag{5.19}$$

with

$$Z_3^{-1} := \lim_{s_0 \to 0} \left[1 + \frac{e_0^2}{12\pi^2} \int_{s_0}^\infty \frac{ds}{s} e^{-im^2 s} \right] \tag{5.20}$$

Here we subtracted the divergent term and added it again in order to combine it with $L^{(0)}$. If we now introduce with

$$\begin{aligned}
B &= B_0 Z_3^{-1/2} \\
e &= e_0 Z_3^{+1/2}
\end{aligned} \tag{5.21}$$

the renormalized field strength B, and charge e, it follows because of $e_0 B_0 = eB$ from (5.19)

$$\mathcal{L} = -\frac{1}{2}B^2 + \frac{1}{8\pi^2}\int_0^\infty \frac{ds}{s^3}\, e^{-im^2 s}\left[(eBs)\cot(eBs) + \frac{1}{3}(eBs)^2 - 1\right]$$

$$= \mathcal{L}_R^{(0)} + \mathcal{L}_R^{(1)} \qquad (5.22)$$

with the gauge invariant renormalized Lagrangian

$$\mathcal{L}_R^{(0)} = -\frac{1}{2}B^2$$

and

$$\mathcal{L}_R^{(1)} = \frac{1}{8\pi^2}\int_0^\infty \frac{ds}{s^3}\, e^{-im^2 s}\left[(eBs)\cot(eBs) + \frac{1}{3}(eBs)^2 - 1\right] \qquad (5.23)$$

That $L^{(1)}$ really is finite becomes clear if one expands the integrands

$$\frac{1}{s^3}\left[(eBs)\cot(eBs) + \frac{1}{3}(eBs) - 1\right] = -\frac{1}{45}(eB)^4 s + \cdots \qquad (5.24)$$

So it is decisive for the renormalization of L that the divergent part of $L^{(1)}$ has exactly the structure of $L^{(0)}$, i.e., is quadratic in B. The fact that the mass need not be renormalized is a special characteristic of the one-loop approximation which, as we shall see, does not apply to $L^{(2)}$.

We can get another equivalent form of $L_R^{(1)}$ from (5.23) by rotating the integration path by $s \to -is$ according to the $i\varepsilon$ rule $(m^2 \to m^2 - i\varepsilon)$, which because of $\cot(ix) = -i\coth(x)$ leads to

$$\mathcal{L}_R^{(1)}(B) = -\frac{1}{8\pi^2}\int_0^\infty \frac{ds}{s^3}\, e^{-m^2 s}\left[(eBs)\coth(eBs) - \frac{1}{3}(eBs)^2 - 1\right] \qquad (5.25)$$

It becomes clear in this representation that for purely magnetic fields of arbitrary strength, Im $L^{(1)}(B)$ and thus the pair production probability disappears. This is a result of the fact that a purely magnetic field cannot transfer energy to a charged particle (in our case to the virtual electron and positrons).

In all our previous calculations we have always presumed a pure magnet field; however, the contrary limiting case of a pure electric field can easily be derived from this by considering that L as a gauge invariant Lorentz scalar, must be written as a function of the only two gauge invariant Lorentz scalars of the Maxwell field

$$\mathcal{F} = \frac{1}{4} F_{\mu\nu} F^{\mu\nu} = \frac{1}{2}(\vec{B}^2 - \vec{E}^2)$$

$$\mathcal{G}^2 = \left(\frac{1}{4} F_{\mu\nu} {}^*F^{\mu\nu}\right)^2 = (\vec{E}\cdot\vec{B})^2, \quad {}^*F_{\mu\nu} = \frac{1}{2}\varepsilon_{\mu\nu\alpha\beta} F^{\alpha\beta}$$

The conversion from pure electric to pure magnetic field takes place by the substitution $B \to \frac{1}{i} E$ since G^2 vanishes in both cases and the necessary change in the sign of F is caused by exactly this substitution. So, from (5.25) we get

$$\mathcal{L}_R^{(1)}(E) = -\frac{1}{8\pi^2}\int_0^\infty \frac{ds}{s^3}\, e^{-m^2 s}\left[(eEs)\cot(eEs) + \frac{1}{3}(eEs)^2 - 1\right] \tag{5.26}$$

This is exactly Schwinger's [3] original result for the special case of a constant electric and vanishing magnetic field. Since the integrand in (5.26) has poles on the real axis, $\operatorname{Im} L_R^{(1)}(E) \neq 0$, and we get a non-vanishing pair production probability [3,9].

The integral in (5.25) was explicitly calculated by Dittrich [8] by dimensional regularization; the result is

$$\mathcal{L}_R^{(1)}(B) = -\frac{1}{32\pi^2}\Big\{(2m^4 - 4m^2(eB) + \tfrac{4}{3}(eB)^2)\Big[1 + \ln\Big(\frac{m^2}{2eB}\Big)\Big]$$

$$+ 4m^2(eB) - 3m^4 - (4eB)^2\,\zeta'\Big(-1, \frac{m^2}{2eB}\Big) \qquad (5.27)$$

Here ζ represents the Riemann Zeta function in two arguments. (An equivalent form is given in [40]). We shall derive this result in the next section using a completely different method. It remains to be observed that according to [9], eq. (5.27) can also be written in the form

$$\mathcal{L}_R^{(1)}(B) = -\frac{1}{32\pi^2}\Big\{-3m^4 + 4(eB)^2\big(\tfrac{1}{3} - 4\zeta'(-1)\big) + 4m^2(eB)(\ln 2\pi - 1)$$

$$- 2m^4 \ln\frac{2eB}{m^2} - 4m^2(eB)\ln\frac{2eB}{m^2} - \frac{4}{3}(eB)^2 \ln\frac{2eB}{m^2}$$

$$- 16(eB)^2 \int_1^{1+\frac{m}{2eB}} dx\, \ln\Gamma(x)\Big\} \qquad (5.28)$$

whereby a numerical evaluation of $L_R^{(1)}$ became possible [9,10,11]. We want to use (5.28) to calculate $L_R^{(1)}$ in the limiting case of strong fields, i.e., for $(eB)/m^2 \gg 1$. First we show that for those values of the field strength, the integral over the logarithm of the gamma function only yields a constant [9] $(b = (eB)/m^2)$:

$$b^2 \int_1^{1+\frac{1}{2b}} dx\, \ln\Gamma(x) \approx b^2 \int_1^{1+\frac{1}{2b}} dx \Big[\ln\Gamma(1) + \frac{d}{dx}\ln\Gamma(x)\Big|_{x=1}(x-1)\Big]$$

$$= b^2\, \psi(1) \int_1^{1+\frac{1}{2b}} dx\, (x-1)$$

$$= \frac{1}{8}\psi(1) = -\frac{1}{8}C$$

Here, C is Euler's number. By only considering the dominant terms for large field strength from (5.28), we get the asymptotic form of the one-loop effective Lagrangian

$$\mathcal{L}_R^{(1)}(B) \approx -\frac{1}{32\pi^2}\left[4(eB)^2\left(\frac{1}{3}-4\zeta'(-1)\right)-\frac{4}{3}(eB)^2\ln\frac{2eB}{m^2}\right]$$

or, with $\alpha = e^2/4\pi$:

$$\mathcal{L}_R^{(1)}(B) \approx \frac{\alpha B^2}{6\pi}\left[\ln\frac{eB}{m^2}+12\zeta'(-1)-1+\ln 2\right] \qquad (5.29)$$

We shall come back to this expression in the next section when we examine the Lagrangian for massless spinor QED.

Next, let us look how $L_R^{(1)}$ behaves as a function of B. Equation (5.27) has been evaluated numerically by several authors [10,11]. It turns out that $L_R^{(1)}$ is a monotonically increasing function of B for all $B > 0$ (this can also be shown analytically [11]). If we plot $L_R^{(1)}$ against B in units of the "critical field strength" $B_{cr} \equiv m^2/e \equiv 4.4 \cdot 10^{13}$ Gauss, we get the following diagram*

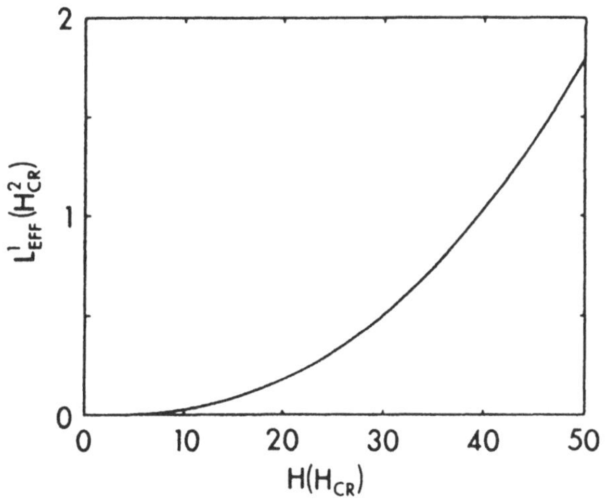

*All diagrams are taken from ref. [11].

Hereby $L_R^{(1)}$ is given in units of B_{cr}^2. Note that in order to obtain the complete effective Lagrangian $L_{eff} = L^{(o)} + L^{(1)} + \ldots$ one has to add the dominating classical contribution $-1/2\ B^2$ to the above diagram. The sum is a monotonically decreasing function B.

Making the substitution $B \to -i\ E$ in (5.27), one gets a complex function for $L_R^{(1)}(E)$. Its real part has the following behaviour for $0 < E < 50\ E_{cr}$ with $E_{cr} \equiv m^2/e = 1.7 \cdot 10^{16}$ V/cm being the electrical critical field strength:

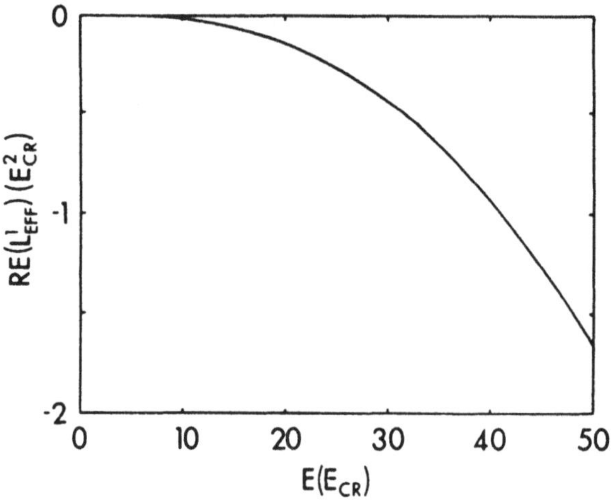

It is interesting that Re $L_R^{(1)}(E)$ possesses a maximum at about $E \approx 3\ E_{cr}$, which is not resolved in the above plot; investigating the range $0 < E < 5\ E_{cr}$ reveals the following structure for the negative real part of $L_R^{(1)}(E)$:

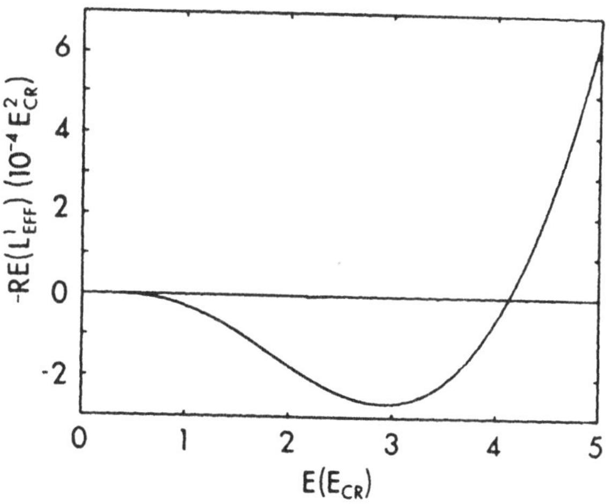

Note that $L_R^{(1)}$ now is measured in units of $10^{-4}\ E_{cr}^2$. Again, to obtain the total effective Lagrangian, one has to add the Maxwellian contribution $1/2\ E^2$ to the above values. Then, just as in the magnetic case, the complete Lagrangian is a monotonic function of E because the quantum corrections $L_R^{(1)}$ are more than compensated for by $1/2\ E^2$.

Thus it is shown that (at least within the considered range for E and B) $L_{eff} = L_R^{(o)} + L_R^{(1)}$ is monotonic in E and B, respectively, and its only extremum is at E = 0 and B = 0. This in turn implies that the effective potential V_{eff} calculated from L_{eff} via the usual Legendre transform [37]

$$V_{eff}(\vec{E},\vec{B}) = \sum_i E_i \frac{\partial \mathcal{L}_{eff}}{\partial E_i} - \mathcal{L}_{eff}(\vec{E},\vec{B}) \tag{5.30}$$

has a unique minimum for E = 0 and B = 0, respectively. These findings are of some importance, because they show that the phenomenon of spontaneous symmetry breaking [51,53] does not occur in a pure-electric- or pure-magnetic-field case in quantum electrodynamics.

This means the following: The true ground-state, i.e., true vacuum, of a field theory is defined as that state $|0\rangle$ which minimizes the expectation value of the Hamiltonian, i.e., it fulfils (we assume $\langle 0|0\rangle = 1$).

$$\langle 0|H|0\rangle \leq \langle \psi|H|\psi\rangle \tag{5.31}$$

for all $|\psi\rangle$ with $\langle\psi|\psi\rangle = 1$. Determining the vacuum state according to (5.31), two cases can occur. Either the vacuum is uniquely given by (5.31) or there are several states $|0\rangle$ satisfying this minimalization criterion. The latter case is referred to as vacuum degeneracy. Until 1960 one only discussed those relativistic quantum field theories in which the vacuum state is not degenerate. (To be precise, one assumed this not to be the case.) However, since the work of Nambu and Goldstone [55], we know that as a resultat of spontaneous symmetry breakdown, such a degeneracy can occur also in relativistic theories. Let us illustrate this point by a simple example. Consider the theory of a complex scalar field defined by the Lagrangian

$$\mathcal{L} = -\partial_\mu \phi^* \partial^\mu \phi - V(\phi) \tag{5.32}$$

with

$$V(\phi) = -\tfrac{1}{2}\mu^2 \phi^*\phi + \tfrac{1}{2}\lambda^2(\phi^*\phi)^2 \tag{5.33}$$

and μ and λ being two real constants. Now let us try to find the ground state of this model in the lowest ("tree") approximation. To this end we study the vacuum expectation value of the field operator $\phi(x)$:

$$\phi_{cl}(x) = \langle 0|\phi(x)|0\rangle \tag{5.34}$$

Obviously, ϕ_{cl} is a classical (c-number) field. If the ground-state is assumed to be translational invariant (which is usually done), we must have

$$\phi_{cl}(x) = const \equiv \phi_{cl} \in \mathbb{C} \quad ,$$

i.e., $\phi_{cl}(x)$ must be independent of the space-time point x. The magnitude of the constant ϕ_{cl} remains to be determined. To lowest order, i.e., when radiative correction are neglected, it is given by the value minimizing the classical potential (5.33):

$$\left.\frac{\partial V}{\partial \phi}\right|_{\phi=\phi_{cl}} = 0$$

This leads to

$$|\phi_{cl}|^2 \equiv |\phi_{cl}(x)|^2 = \frac{\mu^2}{2\lambda}$$

or

$$\phi_{cl}(x) = \sqrt{\frac{\mu^2}{2\lambda}}\, e^{i\varphi} \tag{5.35}$$

with an <u>arbitrary</u> phase angle $\varphi \in [0,2\pi)$. We see that one does not get a unique ground-state, but an infinite number of vacua parametrized by the angle φ. This still can be considered from a different point of view. It is easy to see that (5.32) with (5.33) is invariant under the phase transformation

$$\phi(x) \longrightarrow e^{i\alpha}\phi(x) \tag{5.36}$$

with a constant $\alpha \in \mathbb{R}$. Due to Noether's theorem, this symmetry of L gives rise to the conservation of the quantum number carried by the complex ϕ-field. However, returning to (5.35), it is clear that a given vacuum state is characterized by <u>one</u> value of φ, i.e., by multiplying (5.35) with exp(iα), we come from one vacuum to another; thus, given the theory based on a given vacuum, this vacuum is not invariant under the transformations (5.36)!

This is a simple example of a theory where a symmetry of the Lagrangian is not shared by the ground-state due to a vacuum degeneracy.
At first sight, it appears that such a phenomenon can not occur in electrodynamics because the fields $\vec{E}$ and $\vec{B}$ (or A_μ) are vectors and, by acquiring a non-vanishing vacuum expectation value, they would single out a direction and thus Lorentz invariance of the vacuum would be broken. However, one could conceive of a kind of domain structure for the vacuum which restores Lorentz invariance at a macroscopic level, but exhibits microscopic "bubbles" with randomly oriented fields. This would be similar to the domain structure of ferromagnetica. (Such a vacuum was indeed proposed for quantum chromodynamics; see section (9)).
Now, our results of the above calculations indicate that such a symmetry breaking should not occur in QED up to the one loop level. At the classical level, of course, electrodynamics provides a unique solution for the minimalization of $H = \int d^3x \, \frac{1}{2}(E^2+B^2)$, viz. E=B=0, and so a vacuum degeneracy, if any, would have to arise as a quantum effect.

The situation in QED, where the vacuum seems to have a vanishing expectation value for $\vec{E}$ and $\vec{B}$, has to be contrasted with QCD, where, as we shall discuss in section (9), one finds

$\langle 0|F^a_{\mu\nu}\ F^a_{\mu\nu}|0\rangle \neq 0.$

After now having discussed the properties of the real part of $L_R^{(1)}$, let us turn to its imaginary part, which, via $w = 2\text{Im}\ L$, is responsible for the pair creation of an external field. First let us consider the case of a pure magnetic field. Looking at eq. (5.25), we see that the integrand in the integral for $L_R^{(1)}(B)$ is real and has no singularities on the path of integration. This means that $L_R^{(1)}(B)$ is purely real and so $w = 2\ \text{Im}\ L_R^{(1)}$ vanishes. The fact that a magnetic field does not produce pairs can be understood in simple physical terms. Because the Lorentz force is always perpendicular to the velocity, it does not transfer energy from the field to a particle; hence the field can not supply the energy $2mc^2$ necessary for a virtual electron pair to become a real one.

To deal with the electric field, we return to (5.23) and substitue $E \rightarrow -iB$; the result is

$$\mathcal{L}_R^{(1)}(E) = \frac{1}{8\pi^2}\int_0^\infty \frac{ds}{s^3}\, e^{-im^2 s}\Big[(eEs)\coth(eEs) - \frac{1}{3}(eEs)^2 - 1\Big] \tag{5.37}$$

The imaginary part of this integral is easily calculated using the method of the residues. First we take advantage of the fact that the imaginary part of

$$e^{-im^2 s} = \cos(m^2 s) - i\sin(m^2 s)$$

is an odd function of s, whereas the real part is even. Because s^{-3} times the square bracket in (5.37) is also odd, we have

$$\mathrm{Im}\,\mathcal{L}_R^{(1)}(E) = \frac{1}{2}\,\frac{1}{i}\,\frac{1}{8\pi^2}\int_{-\infty}^{\infty}\frac{ds}{s^3}\,e^{-im^2 s}\,\Big[(eEs)\coth(eEs) - \frac{1}{3}(eEs)^2 - 1\Big] \tag{5.38}$$

with the integration contour now being the whole real axis of the complex s-plane. Note that the integrand has poles on the imaginary axis due to the coth-function:

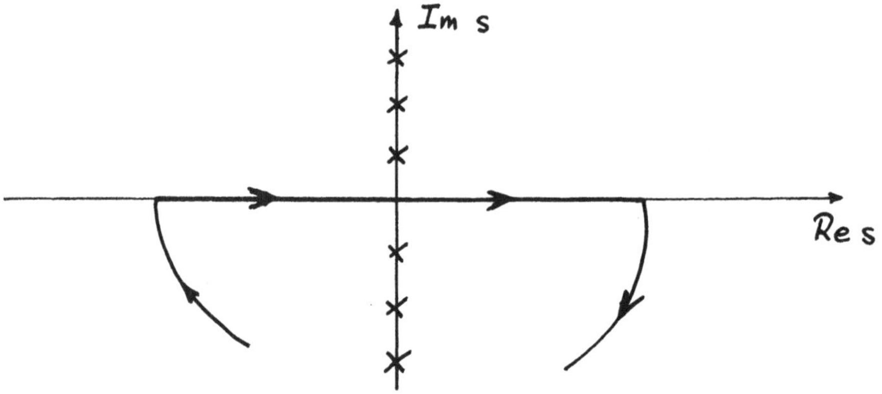

(The crosses denote the positions of the poles).
Without altering the integral we may close the contour by a semi-circle in the lower half-plane; hence Im $L_R^{(1)}$ is given by $(-2\pi i)$ times the sum of the residues of the poles on the negative imaginary axis:

$$\mathrm{Im}\,\mathcal{L}_R^{(1)}(E) = \frac{1}{2i}(-2\pi i)\sum_{n=1}^{\infty}\mathrm{Res}\Big(\frac{1}{8\pi^2}\frac{1}{s^3}\,e^{-im^2 s}\times \times\Big[(eEs)\coth(eEs) - \cdots\Big]\ ;\ s = -i\frac{n\pi}{eE}\Big) \tag{5.39}$$

(The same result would also be obtained by starting from (5.26) and choosing an integration contour which passes the poles due to the cot-function in the upper half-plane). Recalling that $\coth(ax) = 1/ax + \ldots$, we immediately obtain

$$\mathrm{Im}\,\mathcal{L}_R^{(1)}(E) = \frac{1}{8\pi}\sum_{n=1}^{\infty}\left(\frac{eE}{n\pi}\right)^2 e^{-\frac{m^2 n\pi}{eE}}$$

$$= \frac{\alpha}{2\pi^2}E^2\sum_{n=1}^{\infty}\frac{1}{n^2}e^{-\frac{m^2 n\pi}{eE}} \qquad (5.40)$$

This sum can not be done in closed form; nevertheless, it is easy to get simple expressions for the limiting cases of very strong or very weak fields. For strong fields ($eE \gg m^2$), the exponential in (5.40) is close to unity and exploiting [36]

$$\sum_{n=1}^{\infty} n^{-2} = \zeta(2) = \frac{\pi^2}{6}$$

one obtains

$$\mathrm{Im}\,\mathcal{L}_R^{(1)}(eE \gg m^2) = \frac{(eE)^2}{48\pi} = \frac{\alpha}{12}E^2 \qquad (5.41)$$

This result could also have been derived from (5.29) by substituting $E \to -iB$.

For very weak fields ($eE \ll m^2$), the contributions of the $n = 2,3\ldots$ terms in (5.40) can be negelected and one ends up with

$$\mathrm{Im}\,\mathcal{L}_R^{(1)}(eE \ll m^2) = \frac{1}{8\pi^3}(eE)^2 e^{-\frac{m^2\pi}{eE}} \qquad (5.42)$$

For intermediate values of E, numerical methods are necessary; one then gets the following plot [11]

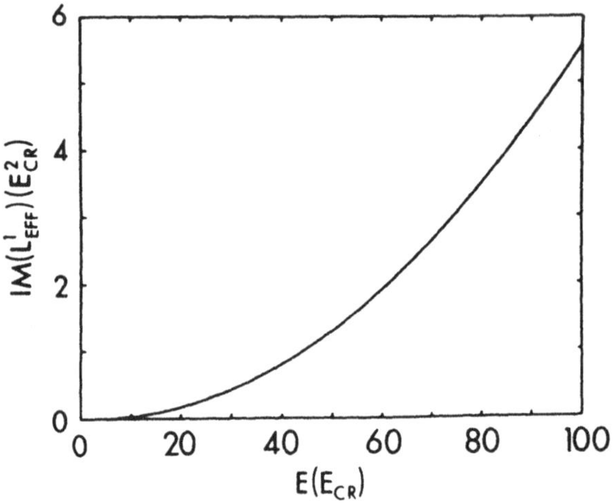

The values are again expressed in units of $E_{cr} \equiv m^2/e = 1.7 \cdot 10^{16}$ V/cm.

Now that we have explicit formulae for the pair creation probability $2\,\mathrm{Im}\,L_R^{(1)}$, it is natural to ask for which values of E pair creation becomes significant. The region where such effects become observable is reached for fields with about $10^{-2}\,E_{cr} \approx 10^{14}$ V/cm. However, this is by many orders of magnitude away from the field strengths which can be produced by experimentalists. These have a magnitude of, say, 10^8 V/cm, giving a rate for the pair production of about $10^{-(10^8)}$ (!) which is far from being experimentally observable.

Nevertheless, in nature there are field strengths with magnitudes near E_{cr}, namely at the surface of a heavy nucleus. In this case, however, the strong interactions dominate and the QED effects are at best small corrections. (Furthermore, the above results are, strictly speaking, valid only for constant fields. But for non-constant fields, the order of magnitude of Im L is presumably not very different from that of constant fields).

Finally let us look a little closer at the functional form of (5.42). Due to the appearance of the factor 1/e in the argument of the exponential, Im $L_R^{(1)}(eE << m^2)$ is a non-analytic function in e. (The same is true for every individual term in the sum (5.40)). This means that Im $L^{(1)}$ vanishes in every finite order of perturbation theory around e = 0. In computing Im $L^{(1)}$ we did a geniunely non-perturbative calculation, because the functional $W^{(1)}$ represents the sum of an infinite number of Feynman diagrams. (Cf. appendix G).

Another interesting point is revealed by reinstating $\hbar$ (and c) in (5.42); the exponential then becomes

$$\exp\left[-\frac{m^2c^3}{e\hbar}\frac{\pi}{E}\right] \tag{5.43}$$

Obviously, Im L is also non-analytical in $\hbar$. This means that pair creation cannot be computed by starting from a classical configuration and then calculating small quantum fluctuations (of order $\hbar$, say) around this configuration. So pair production is an intrinsically qunatum mechanical process, which is not seen in any finite order of $\hbar$. This situation is similar to that of a particle with energy E which tunnels through a potential barrier V(x); the amplitude for a tunneling process to occur is given by the Gamow factor

$$\exp\left[-\frac{2}{\hbar}\int dx\,(2m\{V(x)-E\})^{1/2}\right]$$

featuring the same non-analytic $\hbar$-dependence as (5.43). Hence tunneling does not occur in every finite order of an expansion in $\hbar$.

The notion of pair creation being a tunneling process can be made more precise by using Dirac's hole theory. For $\vec{E} = 0$, the vacuum is characterized by a completely filled lower Dirac sea, whereas the upper is completely empty:

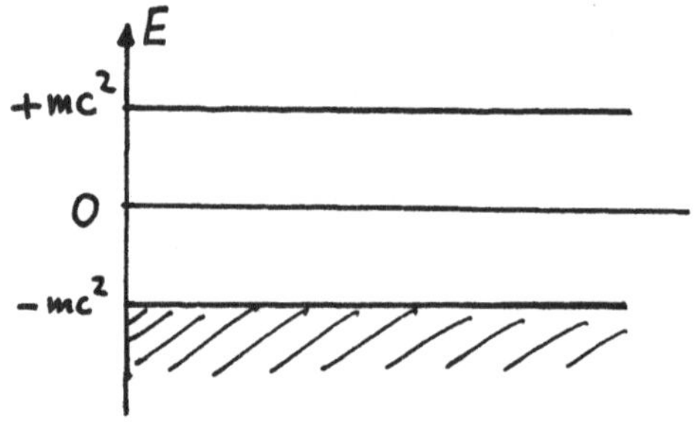

Now we switch on a constant electric field $\vec{E} = |\vec{E}|\ \hat{e}_x$ directed along the positive x-axis; the corresponding potential is $A^0 = -|\vec{E}|x$ and the electrostatic energy for an electron is $W = -eA^0 = e|\vec{E}|x$ $(e > 0)$. Hence we get the following shift of the energy levels (as a function of the spatial coordinate x):

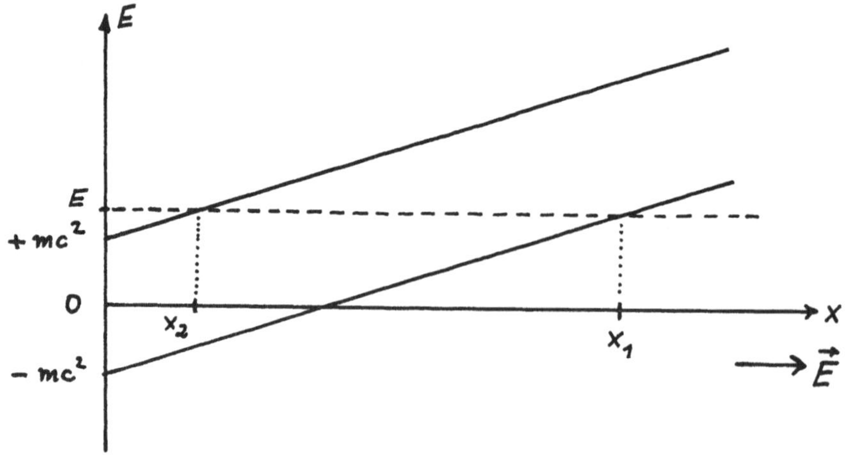

Consider an electron of the lower Dirac sea sitting at x_1 with energy E; if it succeeds in tunneling through the forbidden region between x_1 and x_2 it may occupy a level of the upper Dirac sea at x_2 and would then be accelerated by $\vec{E}$ in the negative x-direction. When leaving x_1, however, it leaves behind a hole, which will be accelerated in the opposite direction. It is in this sense that e^+/e^- creation can be considered as a tunneling process. Note that if $|\vec{E}|$ increases, the distance $|x_2-x_1|$ decreases and hence the tunneling probability gets larger.

Up to now we have always worked either with a pure electric or a pure magnetic field. For the sake of completeness we finally also write down $L_R^{(1)}$ for the case when electric and magnetic fields are present simultaneously:

$$\mathcal{L}_R^{(1)}(\vec{E},\vec{B}) = \frac{1}{8\pi^2}\int_0^\infty \frac{ds}{s^3}\, e^{-ism^2}\Big[e^2s^2\,\mathcal{G}\cdot \cot\left[es\left(\sqrt{\mathcal{F}^2+\mathcal{G}^2}+\mathcal{F}\right)^{\frac{1}{2}}\right]\coth\left[es\left(\sqrt{\mathcal{F}^2+\mathcal{G}^2}-\mathcal{F}\right)^{\frac{1}{2}}\right] - 1 + \frac{2}{3}e^2s^2\,\mathcal{F}\Big] \tag{5.44}$$

with F and G given above.

For a derivation using the elegant proper-time method, the reader is referred to Schwinger's paper [3]. We see that it is indeed possible to write $L_R^{(1)}$, and hence $W^{(1)}$, entirely in terms of the gauge invariant Lorentz scalars F and G^2. According to our discussion in section (1), this gauge invariance of $W^{(1)}$ assures the consistence of the generalized Maxwell equations (1.6). Eq. (5.44) could serve as a starting point for the actual computation of the non-linear effects mentioned in section (1). This task is complicated by the fact that the integral (5.44) cannot be solved in closed form. For the application to the photon splitting process, for instance, see Adler [2].

With these remarks we terminate our discussion of $L_R^{(1)}$ and turn in the next section to the promised derivation of the explicit representation (5.27).

Remark:

Our result (5.29) is identical with Ritus' [4]

$$\mathcal{L}_R^{(1)}(B) \approx \frac{\alpha B^2}{6\pi}\left[\ln \frac{eB}{\gamma\pi m} + \frac{6}{\pi^2}\zeta'(2)\right]$$

As proof, one uses the functional equation [36],

$$2^z\,\Gamma(1-z)\,\zeta(1-z)\,\sin\frac{z\pi}{2} = \pi^{1-z}\,\zeta(z) \;,$$

differentiates with respect to z and then sets z = -1. After substituting the easily obtainable values [36] for $\zeta(2)$, $\zeta(-1)$ and $\Gamma'(2)$ one then gets the desired relation between $\zeta'(-1)$ and $\zeta'(2)$.

(6) The Zeta Function

In this section the concept of the ζ-function regularization [12,15] will be applied to the one-loop effective Lagrangian of spinor QED; thereby we shall show that these methods produce exactly eq. (5.27), but with much less calculatory effort than by dimensional regularization.

We begin directly at

$$i W^{(1)}[A] = - \operatorname{Tr} \ln G_+[A] \tag{6.1}$$

where, as will be seen, we can do without the additional normalization factor $G_+[0]$. This can also be written as

$$i W^{(1)}[A] = - \ln \det G_+[A] \tag{6.2}$$

so that one now must calculate the determinant of the propagator. The usual definition of 'det A' as product of the eigenvalue of an operator A is unfitting here, however, since this product diverges in our case. We will see that one can get the correct generalization of the determinant definition by considering the following:

One considers an operator A with positive, real, discrete eigenvalues $\{a_n\}$, i.e. $Af_n(x) = a_n f(x)$ is valid and one defines its associated Zeta function by

$$\zeta_A(s) := \sum_n a_n^{-s} = \sum_n e^{-s \ln a_n} \tag{6.3}$$

where n runs over all eigenvalues. If one chooses for A the Hamilton operator of the harmonic oscillator for example, then one gets (apart from the zero point energy) exactly the Riemann Zeta function. By formal differentiation, now follows:

$$\zeta_A'(0) = -\sum_n \ln a_n \, e^{-s \ln a_n}\Big|_{s=0}$$

$$= -\ln\big(\prod_n a_n\big)$$

This suggests the definition

$$\det A := \exp[-\zeta_A'(0)] \tag{6.4}$$

which we shall exclusively be using in the following. The advantage of this method is that $\zeta_A'(0)$ is not singular for many operators of physical interest .

Now we bring W[A] into a form which leads only to differential operators of the second order, allowing a simple calculation of the associated Zeta function later on; first, it follows from (6.1):

$$\begin{aligned} W^{(1)}[A] &= i \, \mathrm{Tr} \ln G_+[A] \\ &= -i \ln \det(m + \not{\pi} - i\varepsilon) \end{aligned} \tag{6.5}$$

Now we use

$$-\not{\pi}^2 = \pi^2 - \frac{e}{2}\sigma_{\mu\nu}F^{\mu\nu} = \pi^2 - eB\sigma^3$$

and get

$$\begin{aligned} &\mathrm{Tr}\ln(m+\not{\pi}) + \mathrm{Tr}\ln(m-\not{\pi}) \\ &= \mathrm{Tr}\ln(m^2 - \not{\pi}^2) \\ &= \mathrm{Tr}\ln\left[(m^2+\pi^2)I_4 - eB\sigma^3\right] \end{aligned} \tag{6.6}$$

If we denote the trace in spinor (coordinate) space as tr_γ (tr_x) and the i-th eigenvalue of the matrix A as $EW_i(A)$, then

$$\mathrm{Tr}\,\ln\left[(m^2+\pi^2)I_4 - eB\sigma^3\right]$$

$$\equiv \mathrm{tr}_x\,\mathrm{tr}_\gamma\,\ln\left[(m^2+\pi^2)I_4 - eB\sigma^3\right]$$

$$= \mathrm{tr}_x \sum_{i=1}^{4} \ln EW_i \begin{bmatrix} m^2+\pi^2-eB & & & 0 \\ & m^2+\pi^2+eB & & \\ & & m^2+\pi^2-eB & \\ 0 & & & m^2+\pi^2+eB \end{bmatrix}$$

$$= 2\,\mathrm{tr}_x\left[\ln(m^2+\pi^2-eB) + \ln(m^2+\pi^2+eB)\right]$$

Substitution into (6.6) gives then

$$\mathrm{Tr}\,\ln(m+\not{\pi}) + \mathrm{Tr}\,\ln(m-\not{\pi})$$

$$= 2\,\mathrm{tr}_x\left[\ln(m^2+\pi^2-eB) + \ln(m^2+\pi^2+eB)\right]$$

or, with $\det_{\gamma x} \equiv \det$:

$$\ln\det_{\gamma x}(m+\not{\pi}) + \ln\det_{\gamma x}(m-\not{\pi})$$
$$= 2\ln\det_x(m^2+\pi^2-eB) + 2\ln\det_x(m^2+\pi^2+eB) \tag{6.7}$$

But the determinant of $(m \pm \not{\pi})$ is a Lorentz scalar which means in particular that it does not depend on the sign of the $\not{\pi}$, i.e. it is

$$\det_{\gamma x}(m+\not{\pi}) = \det_{\gamma x}(m-\not{\pi}) .$$

From (6.7) it follows

$$\ln\det_{\gamma x}(m+\not{\pi}) = \ln\det_x(m^2+\pi^2+eB)$$
$$+ \ln\det_x(m^2+\pi^2-eB) \tag{6.8}$$

which, put into (6.5), gives:

$$W^{(1)}[A] = -i\left[\ln\det_x(m^2+\pi^2+eB)+\ln\det_x(m^2+\pi^2-eB)\right] \quad (6.9)$$

This equation is, however, not correct regarding the dimensions, since in (6.1) we left out the $G_+[0]$ in the denominator, the right side of (6.8) now has the dimension $(\mathrm{mass})^2$, while the left side in our units ($\hbar = c = 1$) is dimensionless. Thus, we introduce an arbitrary parameter μ with the dimension of a mass and replace (6.9) by

$$\begin{aligned} W^{(1)}[A] = -i\,\{&\ln\det_x[\mu^{-2}(m^2+\pi^2+eB)] \\ &+\ln\det_x[\mu^{-2}(m^2+\pi^2-eB)]\} \end{aligned} \quad (6.10)$$

With the determinant definition (6.4) becomes

$$\begin{aligned} W^{(1)}[A] = -i\,\{&\ln\exp[-\zeta'_{\mu^{-2}(m^2+\pi^2+eB)}(0)] \\ &+\ln\exp[-\zeta'_{\mu^{-2}(m^2+\pi^2-eB)}(0)]\} \\ = {}& i\,\zeta_2'(0) \end{aligned} \quad (6.11)$$

with

$$\zeta_2(s) := \zeta_{\mu^{-2}(m^2-i\varepsilon+\pi^2+eB)}(s) + \zeta_{\mu^{-2}(m^2-i\varepsilon+\pi^2-eB)}(s) \quad (6.12)$$

where we again substituted $m^2 - i\varepsilon$ for m^2. Now we perform a Wick rotation by the substitution $t \to \tau = it$ and get for (6.12)

$$\zeta_2(s) = \zeta_{\mu^{-2}(m^2+\pi_E^2+eB)}(s) + \zeta_{\mu^{-2}(m^2+\pi_E^2-eB)}(s) \quad (6.13)$$

with the Euclidean momentum π_E. In order to calculate the zeta function (6.12) or (6.13) according to (6.3) we need the spectrum of the operator

$$\begin{aligned} M^2 &:= m^2 + \pi^2 \\ &= m^2 + \left(\tfrac{1}{i}\partial - eA\right)^2 \\ &= m^2 + \left(\tfrac{1}{i}\partial - eA\right)^2_{\parallel} + \left(\tfrac{1}{i}\partial - eA\right)^2_{\perp} \\ &= m^2 + \partial_t^2 - \partial_3^2 - (\vec{p} - e\vec{A})^2_{\perp} \end{aligned} \tag{6.14}$$

or its Euclidean analogue

$$M_E^2 = m^2 - \partial_\tau^2 - \partial_3^2 - (\vec{p} - e\vec{A})^2_{\perp} \tag{6.15}$$

The spectrum of the operator $(\vec{p}-e\vec{A})_{\perp}^2$ is known from one-particle quantum mechanics; there it is shown [44] that a particle moving in a constant magnetic field $\vec{B} = B\hat{z}$ and having no momentum in z-direction, is described by a Hamilton operator

$$H = \frac{1}{2m}(\vec{p} - e\vec{A})^2_{\perp}$$

having the eigenvalues

$$E_n = \frac{e}{m} B \left(n + \tfrac{1}{2}\right), \quad n \in \mathbb{N}$$

Thus, the last term of the sum of (6.15) has the eigenvalues

$$2(eB)\left(n + \tfrac{1}{2}\right), \quad n \in \mathbb{N} \tag{6.16}$$

If we imagine further that we enclose the field in a very large normalization volume $\Omega = L^4$ of Euclidean space-time, then we can approximate $(-\partial_\tau^2 - \partial_3^2)$ by plane waves with eigenvalues $(k_0^2 + k_3^2)$, k_0, $k_3 \in \mathbb{R}$, and density $(L/2\pi)^2$. All together one then has

$$\text{spectrum}\,[m^2 + \pi_E^2] =$$

$$\{m^2 + k_0^2 + k_3^2 + (2n+1)eB \mid k_0, k_3 \in \mathbb{R},\ n \in \mathbb{N}\} \tag{6.17}$$

In continuum approximation, the sum in (6.3) can be partially replaced by integrals, leading to

$$\zeta_2(s) = \mu^{2s}\Omega \sum_{n=0}^{\infty} \frac{eB}{2\pi} \iint_{-\infty}^{\infty} \frac{dk_0\, dk_3}{(2\pi)^2} \{[m^2 + k_0^2 + k_3^2 + 2(n+1)eB]^{-s} + [m^2 + k_0^2 + k_3^2 + 2neB]^{-s}\} \tag{6.18}$$

Now we shall first evaluate the k-integrations; to do so, we use the formula [36]

$$\int_0^{\infty} dx\, x^{\mu-1}(1+x^2)^{\nu-1} = \frac{1}{2} B\left(\frac{\mu}{2}, 1-\nu-\frac{\mu}{2}\right)$$

$$\mathrm{Re}\,\mu > 0, \quad \mathrm{Re}\left(\nu + \frac{\mu}{2}\right) < 1 \tag{6.19}$$

and evaluate (6.18) for those values of s for which the integrals exist. (B designates the Beta function or the Euler integral of the first kind) Thereafter, ζ_2 can be analytically continued to a meromorphic function of the whole complex plane. First we consider two special cases of (6.19):

$$\int_{-\infty}^{\infty} dx\, (a^2 + x^2)^{-s} = (a^2)^{\frac{1}{2}-s} B\left(\frac{1}{2}, s-\frac{1}{2}\right) \tag{6.20}$$

furthermore,

$$\int_{-\infty}^{\infty} dx\ (a^2+x^2)^{\frac{1}{2}-s} = (a^2)^{1-s}\ B(\tfrac{1}{2}, s-1) \tag{6.21}$$

Applying (6.20) to (6.18) gives

$$\zeta_2(s) = \mu^{2s}\Omega\ B(\tfrac{1}{2}, s-\tfrac{1}{2}) \sum_{n=0}^{\infty} \frac{eB}{(2\pi)^3} \int_{-\infty}^{\infty} dk_3 \left\{ [m^2 + k_3^2 + (n+1)2eB]^{\frac{1}{2}-s} + [m^2 + k_3^2 + 2neB]^{\frac{1}{2}-s} \right\}$$

with (6.21) it follows that

$$\zeta_2(s) = \mu^{2s}\Omega\ B(\tfrac{1}{2}, s-\tfrac{1}{2})\, B(\tfrac{1}{2}, s-1)\, \frac{eB}{(2\pi)^3} \cdot \sum_{n=0}^{\infty} \left\{ [m^2+2(n+1)eB]^{1-s} + [m^2+2neB]^{1-s} \right\} \tag{6.22}$$

If one uses the two functional equations of the Beta function [36]

$$B(x,x) = 2^{1-2x}\ B(\tfrac{1}{2}, x) \tag{6.23}$$

$$B(x,x)\, B(x+\tfrac{1}{2}, x+\tfrac{1}{2}) = \pi\, [2^{4x-1}\, x]^{-1}$$

then one gets

$$B(\tfrac{1}{2}, s-\tfrac{1}{2})\, B(\tfrac{1}{2}, s-1) =$$

$$= 2^{2s-2}\ 2^{2s-3}\ B(s-\tfrac{1}{2}, s-\tfrac{1}{2})\, B(s-1, s-1)$$

$$= 2^{4s-5}\ \pi\, [2^{4(s-1)-1}\,(s-1)]^{-1}$$

$$= \frac{\pi}{s-1}$$

Substitution into (6.22) gives

$$\zeta_2(s) = \mu^{2s}\, \Omega\, \frac{eB}{(2\pi)^3}\, \frac{\pi}{s-1} \left\{ \sum_{n=0}^{\infty} [m^2 + 2eB(n+1)]^{1-s} + \sum_{n=0}^{\infty} [m^2 + 2eBn]^{1-s} \right\}$$

$$= \mu^{2s}\, \Omega\, \frac{eB}{(2\pi)^3}\, \frac{\pi}{s-1}\, (2eB)^{1-s} \,. \tag{6.24}$$

$$\cdot \left\{ \sum_{n=0}^{\infty} \left(\frac{m^2}{2eB} + n + 1\right)^{1-s} + \sum_{n=0}^{\infty} \left(\frac{m^2}{2eB} + n\right)^{1-s} \right\}$$

If one keeps in mind that the Riemann Zeta function in two arguments has the representation [36]

$$\zeta(z,q) = \sum_{n=0}^{\infty} (q+n)^{-z} \,, \quad \mathrm{Re}\, z > 1 \tag{6.25}$$

then it is clear that for the second sum one gets

$$\sum_{n=0}^{\infty} \left(\frac{m^2}{2eB} + n\right)^{1-s} = \zeta\left(s-1, \frac{m^2}{2eB}\right)$$

and for the first

$$\sum_{n=0}^{\infty} \left(\frac{m^2}{2eB} + 1 + n\right)^{1-s} = \sum_{\nu=1}^{\infty} \left(\frac{m^2}{2eB} + \nu\right)^{1-s}$$

$$= \zeta\left(s-1, \frac{m^2}{2eB}\right) - \left(\frac{m^2}{2eB}\right)^{1-s}$$

Our end result for ζ_2 thus reads

$$\zeta_2(s) = \Omega \frac{eB}{8\pi^2} (s-1)^{-1} (2eB)^{1-s} \left[2\zeta(s-1,q) - q^{1-s} \right] \mu^{2s} \tag{6.26}$$

where we set

$$q := \frac{m^2}{2eB} \tag{6.27}$$

The derivative of this equation is

$$\zeta_2'(s) = \Omega \frac{eB}{8\pi^2} \left\{ -(s-1)^{-2} (2eB)^{1-s} \left[2\zeta(s-1,q) - q^{1-s} \right] \mu^{2s} \right.$$

$$- (s-1)^{-1} \ln(2eB)\,(2eB)^{1-s}\,[2\zeta(s-1,q) - q^{1-s}]\,\mu^{2s}$$

$$+ (s-1)^{-1}\,(2eB)^{1-s}\,[2\zeta'(s-1,q) + \ln q \cdot q^{1-s}]\,\mu^{2s}$$

$$+ (s-1)^{-1}\,(2eB)^{1-s}\,[2\zeta(s-1,q) - q^{1-s}]\,\ln\mu^2\,\mu^{2s} \Big\}$$

or, for s = 0:

$$\zeta_2'(0) = \Omega \frac{(eB)^2}{4\pi^2}\Big\{-[2\zeta(-1,q)-q] + \ln(2eB)[2\zeta(-1,q)-q]$$

$$-[2\zeta'(-1,q)+q\ln q] - \ln\mu^2[2\zeta(-1,q)-q]\Big\}$$

$$= \Omega \frac{(eB)^2}{4\pi^2}\Big\{[q-2\zeta(-1,q)]\,(1-\ln(2eB)+\ln\mu^2) \tag{6.28}$$

$$-[2\zeta'(-1,q)+q\ln q]\Big\}$$

To calculate $\zeta(-1,q)$ we need the following property of the Zeta function [36]

$$\zeta(-n,q) = -\frac{B'_{n+2}(q)}{(n+1)(n+2)}\,, \quad n \in \mathbb{N} \tag{6.29}$$

Here, B_n denotes the Bernoulli polynomials. In particular

$$B_3(q) = q^3 - \frac{3}{2}q^2 + \frac{1}{2}q$$

so that one gets

$$\zeta(-1,q) = -\frac{1}{6}B_3'(q) \tag{6.30}$$

$$= -\frac{1}{2}\left(q^2 - q + \frac{1}{6}\right)$$

From (6.28) together with (6.27) it thus follows

$$\zeta_2'(0) = \Omega \frac{(eB)^2}{4\pi^2} \Big\{ \Big[\Big(\frac{m^2}{2eB}\Big)^2 + \frac{1}{6} \Big] \Big(1 - \ln \frac{2eB}{\mu^2}\Big) - \frac{m^2}{2eB} \ln \frac{m^2}{2eB} \tag{6.31}$$

$$- 2\zeta'\Big(-1, \frac{m^2}{2eB}\Big) \Big\}$$

From this, according to (6.11) one can calculate via

$$W^{(1)}[A] = i\,\zeta_2'(0)$$

the effective action and with

$$W^{(1)}[A] = \int d^3x\, dt\, \mathcal{L}^{(1)}$$

$$= \frac{1}{i} \int d^3x\, d\tau\, \mathcal{L}^{(1)}$$

$$= (-i)\, \Omega\, \mathcal{L}^{(1)}$$

the effective Lagrangian

$$\mathcal{L}^{(1)}(B) = -\Omega^{-1}\, \zeta_2'(0)$$

$$= -\frac{(eB)^2}{4\pi^2} \Big\{ \Big(\frac{m^2}{2eB}\Big)^2 + \frac{1}{6} - \Big(\frac{m^2}{2eB}\Big)^2 \ln \frac{2eB}{\mu^2} \tag{6.32}$$

$$- \frac{1}{6} \ln \frac{2eB}{\mu^2} - \frac{m^2}{2eB} \ln \frac{m^2}{2eB} - 2\zeta'\Big(-1, \frac{m^2}{2eB}\Big) \Big\}$$

This expression for $L^{(1)}$ still depends on the arbitrary parameter μ, which can be fixed by additional conditions for L. In the case of massive QED studied here, $\mu=m$ can be chosen as reference mass. With this choice (6.32) can be put except for a constant, in the form

$$\mathcal{L}^{(1)}(B) = -\frac{1}{32\pi^2}\Big\{(2m^4 - 4m^2(eB) + \tfrac{4}{3}(eB)^2)\cdot \Big[1 + \ln\frac{m^2}{2eB}\Big] + 4m^2(eB) - 3m^4 - 16(eB)^2\, \zeta'(-1, \frac{m^2}{2eB})\Big\} \tag{6.33}$$

which can be proven by simply multiplying out eq. (6.33). Apparently (6.33) agrees exactly with the form obtained by dimensional regularization (5.27). In particular this means that the result obtained by Zeta function regularization is finite without additional subtraction of divergent counterterms (as in (5.17) and (5.19) for example) and that it vanishes for B = 0. Thus (6.4) appears as a useful generalization of the concept of determinants for operators like G_+.

A further advantage of this method is that one can very easily specialize the above calculation to the case of vanishing Fermion mass (m=0), while this is impossible in the integral representation (5.25) for $L^{(1)}$ without making the integral divergent on the upper limit. For this special case, one reads from (6.24)

$$\zeta_2(s) = \mu^{2s}\,\Omega\,\frac{(eB)}{(2\pi)^3}\,\frac{\pi}{s-1}\,(2eB)^{1-s}\Big\{\sum_{n=0}^{\infty}(n+1)^{1-s} + \sum_{n=0}^{\infty} n^{1-s}\Big\}$$
$$= \mu^{2s}\,\Omega\,\frac{(eB)}{(2\pi)^2}\,\frac{1}{s-1}\,(2eB)^{1-s}\,\zeta(s-1) \tag{6.34}$$

Here, in the last line, the representation

$$\zeta(z) = \sum_{n=1}^{\infty} n^{-z}\,, \quad \mathrm{Re}\, z > 1$$

of the normal Riemann Zeta function was used. Differentiation of (6.34) gives

$$\zeta_2'(s) = \Omega \frac{eB}{(2\pi)^2} \Big[-(s-1)^{-2} (2eB)^{1-s} \zeta(s-1) \mu^{2s}$$
$$+ (s-1)^{-1} [-\ln 2eB] (2eB)^{1-s} \zeta(s-1) \mu^{2s}$$
$$+ (s-1)^{-1} (2eB)^{1-s} \zeta'(s-1) \mu^{2s}$$
$$+ (s-1)^{-1} (2eB)^{1-s} \zeta(s-1) \ln\mu^2 \cdot \mu^{2s} \Big]$$

Now we set s = 0 again:

$$\zeta_2'(0) = \Omega \frac{eB}{4\pi^2} \Big[-(2eB)\zeta(-1) + (2eB)\zeta(-1)\ln(2eB)$$
$$-(2eB)\zeta'(-1) - (2eB)\zeta(-1)\ln\mu^2 \Big]$$

Since for the Riemann Zeta function in one or two arguments

$$\zeta(z, 1) = \zeta(z)$$

is valid ($z \in \mathbb{C}$ arbitrary), then

$$\zeta(-1) = \zeta(-1, q=1) = -\frac{1}{12}$$

and thus

$$\zeta_2'(0) = -\Omega \frac{(eB)^2}{24\pi^2} \Big[\ln \frac{2eB}{\mu^2} - 1 + 12\zeta'(-1) \Big]$$

Since

$$\mathcal{L}^{(1)} = -\Omega^{-1} \zeta_2'(0)$$

the one-loop effective Lagrangian for QED with massless Fermions is then given by

$$\mathcal{L}^{(1)}_{m=0}(B) = \frac{(eB)^2}{24\pi^2}\left[\ln\frac{2eB}{\mu^2} - 1 + 12\,\zeta'(-1)\right]$$

$$= \frac{\alpha B^2}{6\pi}\left[\ln\frac{2eB}{\mu^2} - 1 + 12\,\zeta'(-1)\right] \qquad (6.35)$$

If we would replace the undetermined parameter μ in the denominator of the logarithm by the electron mass m, then this would be precisely the expression (5.29) for the Lagrangian of the massive theory in the limiting case of strong fields! But since there exists no natural mass scale in the theory with m = 0 with respect to which one can measure the field strengths, it is not possible with the help of the renormalization conditions applied to L [28] to eliminate μ from $L^{(1)}_{m=0}$. Nevertheless, the physical content of (6.35) can not depend on μ, i.e.

$$\mu\frac{d}{d\mu}\mathcal{L}(\mu, \alpha(\mu), B(\mu)) = 0 \qquad (6.36)$$

must be valid; here we have indicated with the arguments that L also has an implicit dependence on μ via α and B, since with the choice of a particular value for μ, one also decides on a special renormalized coupling constant or field strength which as we have seen in the case of the massive theory, exactly coincides with the physical coupling constants or field strengths for μ=m [31,32].

With (6.36) we are led to the so-called renormalization group equation

$$\left[\mu \frac{\partial}{\partial \mu} + \mu \frac{d\alpha}{d\mu} \frac{\partial}{\partial \alpha} + \mu \frac{dB}{d\mu} \frac{\partial}{\partial B}\right] \mathcal{L}(\mu, \alpha, B) = 0 \quad (6.37)$$

This partial differential equation tells us that an infinitesimal change in μ can be compensated by a corresponding change in α and a rescaling of B. The above equation can be further simplified by utilizing the relationship between the renormalized quantities e, B and the bare e_o, B_o:

$$\begin{aligned} e &= e_o Z_3^{1/2} \\ B &= B_o Z_3^{-1/2} \end{aligned} \quad (6.38)$$

Here we have used the same notation as in section (5), without, however, having had to explicitly perform the given infinite scale transformation for charge and field strength in the framework of the Zeta function regularization. Furthermore, $\alpha = Z_3 \alpha_o$ and $eB = e_o B_o$, so that for the derivatives required in (6.37) we get

$$\frac{d\alpha}{d\mu} = \alpha_o \frac{dZ_3}{d\mu} = \frac{\alpha}{Z_3} \frac{dZ_3}{d\mu} = \alpha \frac{d \ln Z_3}{d\mu}$$

and

$$\frac{dB}{d\mu} = B_o \frac{dZ_3^{-1/2}}{d\mu} = -\frac{1}{2} B_o Z_3^{-3/2} \frac{dZ_3}{d\mu}$$

$$= -\frac{1}{2} B \frac{d \ln Z_3}{d\mu}$$

Here we used the fact that the bare quantities e_o and B_o are naturally independent of the parameter μ which is only introduced during the regularization process. Now we define

$$\beta_\zeta(\alpha) := \mu \frac{d \ln Z_3}{d\mu} \tag{6.39}$$

and get

$$\mu \frac{d\alpha}{d\mu} = \alpha\, \beta_\zeta(\alpha)$$

$$\mu \frac{dB}{d\mu} = -\tfrac{1}{2}\, \beta_\zeta(\alpha)\, B$$

which, together with (6.37) gives

$$\left[\mu \frac{\partial}{\partial\mu} + \alpha\beta_\zeta(\alpha)\frac{\partial}{\partial\alpha} - \tfrac{1}{2}\beta_\zeta(\alpha)\, B\frac{\partial}{\partial B}\right] \mathcal{L}(\mu,\alpha,B) = 0 \tag{6.40}$$

Note that this equation contains only renormalized quantities. By now substituting the known one-loop approximation for L, we can directly calculate the function β_ζ. From

$$\mathcal{L}(\mu,\alpha,B) = -\tfrac{1}{2}B^2 + \frac{\alpha B^2}{6\pi}\ln\frac{2\sqrt{4\pi\alpha}\,B}{\mu^2} + \frac{\alpha B^2}{6\pi}C$$

with

$$C = -1 + 12\,\zeta'(-1)$$

it follows

$$\left[\mu\frac{\partial}{\partial\mu} + \alpha\beta_\zeta(\alpha)\frac{\partial}{\partial\alpha} - \tfrac{1}{2}\beta_\zeta B\frac{\partial}{\partial B}\right]\mathcal{L}(\mu,\alpha,B)$$

$$= -\frac{\alpha B^2}{3\pi}$$

$$+ \beta_\zeta(\alpha)\frac{\alpha B^2}{6\pi}\left[\ln\frac{2eB}{\mu^2} + C\right] + \beta_\zeta(\alpha)\frac{\alpha B^2}{12\pi^2}$$

$$+ \tfrac{1}{2}\beta_\zeta(\alpha)B^2 - \beta_\zeta(\alpha)\frac{\alpha B^2}{6\pi}\left[\ln\frac{2eB}{\mu^2} + C\right]$$

$$- \beta_\zeta(\alpha)\frac{\alpha B^2}{12\pi^2}$$

$$= \left(\frac{1}{2} \beta_f(\alpha) - \frac{\alpha}{3\pi} \right) B^2$$

$$\stackrel{!}{=} 0$$

The validity of (6.36) implies that

$$\beta_f(\alpha) = \frac{2}{3} \frac{\alpha}{\pi} \tag{6.41}$$

We shall return to this important result in section (8).

Now we want to consider the one-loop effective Lagrangian for the case of finite temperature [18] as well; it will be shown again that the Zeta function regularization reproduces, with relatively little difficulty the results achieved by other methods [19]. The effects of finite temperature can be taken into account [15-17] by performing the substitution in (6.18)

$$\int_{-\infty}^{\infty} \frac{dK_0}{(2\pi)} f(K_0^2) \longrightarrow \sum_{\ell=0}^{\infty} f\left(\left\{ \frac{\pi}{\beta}(2\ell+1) \right\}^2 \right) \tag{6.42}$$

Here, $\beta = 1/kT$ (k = Boltzmann's constant, T = temperature) is a periodicity interval in the Euclidean time $\tau = it$. Then the thermal Zeta function is

$$\zeta_2^\beta(s) = \mu^{2s} \Omega_3 \sum_{n=0}^{\infty} \sum_{\ell=0}^{\infty} \int_{-\infty}^{\infty} \frac{dK_3}{(2\pi)} \Big\{ \Big[m^2 + \left\{ \tfrac{\pi}{\beta}(2\ell+1) \right\}^2 + K_3^2 + 2(n+1)eB \Big]^{-s}$$

$$+ \Big[m^2 + \left\{ \tfrac{\pi}{\beta}(2\ell+1) \right\}^2 + K_3^2 + 2neB \Big]^{-s} \Big\} \cdot \frac{eB}{2\pi}$$

with a three-dimensional normalization volume Ω_3. From this one could caluclate with

$$\mathcal{L}^{(1)}(B,T) = -\Omega^{-1}\,\zeta_2^{\beta\,\prime}(0)$$

$$= -\Omega_3^{-1}\,\beta^{-1}\,\zeta_2^{\beta\,\prime}(0)$$

the Lagrangian of the theory with massive fermions at finite temperature and B $\neq$ 0. Since the integral and the sum in the above formula cannot be represented in closed form, we limit ourselves in the following to a massless QED (m=0) and vanishing external field (B=0). To get the relevant Zeta function for this, we go back to the equation (6.15), which now reduces to $M_E^2 = p_E^2$, where M_E^2 has the spectrum

$$\left\{ K_0^2 + K_1^2 + K_2^2 + K_3^2 \;\middle|\; K_i \in \mathbb{R} \right\}$$

Because of (6.42), then

$$\zeta_2^{\beta}(s) = 2\mu^{2s}\,\Omega_3 \sum_{\ell=0}^{\infty} \iiint_{-\infty}^{\infty} \frac{dK_1\,dK_2\,dK_3}{(2\pi)^3}\left[\left\{\tfrac{\pi}{\beta}(2\ell+1)\right\}^2 + K_1^2 + K_2^2 + K_3^2\right]^{-s}$$

Introduction of polar coordinates gives

$$\zeta_2^{\beta}(s) = 2\mu^{2s}\,\Omega_3 \sum_{\ell=0}^{\infty} \frac{4\pi}{(2\pi)^3} \int_0^{\infty} dK\,K^2 \left[\left\{\tfrac{\pi}{\beta}(2\ell+1)\right\}^2 + K^2\right]^{-s}$$

$$= 8\pi\,\mu^{2s}\,\frac{\Omega_3}{(2\pi)^3} \sum_{\ell=0}^{\infty} \left\{\tfrac{2\pi}{\beta}(\ell+\tfrac{1}{2})\right\}^{3-2s} \int_0^{\infty} dx\cdot x^2\left[1+x^2\right]^{-s}$$

The integration can be performed for suitable values of s with (6.19)

$$\zeta_2^{\beta}(s) = 8\pi\mu^{2s}\,\frac{\Omega_3}{(2\pi)^3}\left(\frac{2\pi}{\beta}\right)^{3-2s} \sum_{\ell=0}^{\infty} (\ell+\tfrac{1}{2})^{3-2s}\;\tfrac{1}{2}\,B(\tfrac{3}{2}, s-\tfrac{3}{2})$$

With (6.25) and the identity

$$B(x,y) = \frac{\Gamma(x)\,\Gamma(y)}{\Gamma(x+y)} \tag{6.43}$$

if follows

$$\zeta_2^\beta(s) = 4\pi\,\mu^{2s}\,\frac{\Omega_3}{(2\pi)^3}\,\zeta(2s-3,\tfrac{1}{2})\,\frac{\Gamma(\frac{3}{2})\,\Gamma(s-\frac{3}{2})}{\Gamma(s)}\left(\frac{2\pi}{\beta}\right)^{3-2s}$$

Since Γ has a pole for $s = 0$,

$$\zeta_2^{\beta\prime}(0) = 4\pi\,\frac{\Omega_3}{(2\pi)^3}\left(\frac{2\pi}{\beta}\right)^3\zeta(-3,\tfrac{1}{2})\,\Gamma(\tfrac{3}{2})\,\Gamma(-\tfrac{3}{2})\,\frac{d}{ds}\frac{1}{\Gamma(s)}\bigg|_{s=0} \tag{6.44}$$

As can be seen from the expansion

$$\Gamma(\varepsilon) = \frac{1}{\varepsilon} + \psi(1) + O(\varepsilon)\,,$$

$$\frac{d}{ds}\,\frac{1}{\Gamma(s)}\bigg|_{s=0} = 1$$

Furthermore, one needs

$$\Gamma(-\tfrac{3}{2}) = \tfrac{4}{3}\sqrt{\pi}$$

and

$$\Gamma(+\tfrac{3}{2}) = \tfrac{1}{2}\sqrt{\pi}$$

Substitution into (6.44) gives

$$\zeta_2^{\beta\prime}(0) = \frac{8}{3}\,\pi^2\,\Omega_3\,\beta^{-3}\,\zeta(-3,\tfrac{1}{2}) \tag{6.45}$$

Now we again use (6.29):

$$\zeta(-3,q) = -\frac{1}{20}\,B_5'(q)$$

By means of the relation [36]

$$B_n'(q) = n\,B_{n-1}(q)$$

this can be transformed into a Bernoulli polynomial of the 4th order

$$\begin{aligned}\zeta(-3,\tfrac{1}{2}) &= -\frac{1}{20}\,5B_4(q)\Big|_{q=\frac{1}{2}}\\ &= -\frac{1}{4}\left(q^4-2q^3+q^2-\frac{1}{30}\right)\Big|_{q=\frac{1}{2}}\\ &= -\frac{1}{8}\,\frac{7}{8\cdot 15}\end{aligned}$$

It follows for (6.15)

$$\zeta_2^{\beta\prime}(0) = -\frac{7}{360}\,\pi^2\Omega_3\,\beta^{-3}$$

So we get the expression for the Lagrangian

$$\begin{aligned}\mathcal{L}^{(1)}_{m=0}(B=0,T) &= -\Omega_3^{-1}\beta^{-1}\,\zeta_2^{\beta\prime}(0)\\ &= \frac{7}{360}\,\pi^2\beta^{-4} \qquad (6.46)\\ &= \frac{7}{360}\,\pi^2 k^4 T^4\end{aligned}$$

which, however, does not yet represent the correct final result. Since from the thermodynamic viewpoint, we are dealing with both an electron and positron gas, each with two spin projection possibilities, an additional factor 2 must be introduced into (6.46):

$$\mathcal{L}^{(1)}_{m=0}(B=0,T) = \frac{2}{3}\,\frac{7}{120}\,\pi^2 k^4 T^4 \qquad (6.47)$$

This is precisely the result given in [19] for the Fermi-Dirac case (i.e., spinor QED). It is interesting that this expression can be written as

$$\mathcal{L}^{(1)}_{m=0}(B=0,T) = \frac{2}{3\pi^2}\int_0^\infty \frac{\varepsilon^2 d\varepsilon}{e^{\varepsilon/kT}+1} \tag{6.48}$$

with the right Fermi distribution. (By way of proof, we use the integral [36]

$$\int_0^\infty \frac{x^{2n-1}dx}{e^{px}+1} = (1-2^{1-2n})\left(\frac{2\pi}{p}\right)^{2n}\frac{|B_{2n}|}{4n}$$

for $n = 2$ and $B_4 = -1/30$.)

In order to have also a system that satisfies Bose-Einstein statistics, we calculate in Appendix B the one-loop effective Lagrangian of scalar QED; then we get for the Zeta function without taking temperature effects into account

$$\zeta_1(s) = \mu^{2s}\Omega\sum_{n=0}^\infty \frac{eB}{2\pi}\iint_{-\infty}^{\infty}\frac{dK_0 dK_3}{(2\pi)^2}\left[K_0^2+K_3^2+m^2+(n+\tfrac{1}{2})(2eB)\right]^{-s} \tag{6.49}$$

From this we get the corresponding equation for $T > 0$ by the substitution [15-17]

$$\int_{-\infty}^{\infty}\frac{dK_0}{(2\pi)} f(K_0^2) \longrightarrow \sum_{\ell=-\infty}^{\infty} f(\{\tfrac{2\pi}{\beta}\ell\}^2)\,,\ \beta = \frac{1}{kT} \tag{6.50}$$

Therefore the corresponding Zeta function is

$$\zeta_1^\beta(s) = \mu^{2s}\Omega_3\sum_{n=0}^\infty\sum_{\ell=-\infty}^\infty \frac{eB}{2\pi}\int_{-\infty}^{\infty}\frac{dK_3}{(2\pi)}\left[\{\tfrac{2\pi}{\beta}\ell\}^2+K_3^2+m^2+(n+\tfrac{1}{2})(2eB)\right]^{-s} \tag{6.51}$$

which again cannot be calculated in closed form for arbitrary values of the mass and field strength. As before, we limit ourselves therefore to $m = 0$ and $B = 0$

$$\zeta_1^\beta(s) = \mu^{2s}\Omega_3\sum_{\ell=-\infty}^\infty\iiint_{-\infty}^{\infty}\frac{dK_1 dK_2 dK_3}{(2\pi)^3}\left[\{\tfrac{2\pi}{\beta}\ell\}^2+\vec{K}^2\right]^{-s} \tag{6.52}$$

Now we again introduce polar coordinates

$$\zeta_1^\beta(s) = \mu^{2s}\Omega_3 \sum_{\ell=-\infty}^{\infty} \frac{4\pi}{(2\pi)^3} \int_0^\infty dk\, K^2\left[\left\{\tfrac{2\pi}{\beta}\ell\right\}^2 + K^2\right]^{-s}$$

Since the integrand is an even function of ℓ, we obtain

$$\zeta_1^\beta(s) = \mu^{2s}\frac{\Omega_3}{(2\pi)^3}\, 2\sum_{\ell=1}^{\infty} 4\pi \int_0^\infty dk\, K^2\left[\left\{\tfrac{2\pi}{\beta}\ell\right\}^2 + K^2\right]^{-s}$$

$$+\mu^{2s}\frac{\Omega_3}{(2\pi)^3}\, 4\pi \int_0^\infty dk\, K^{2-2s} \qquad (6.53)$$

The integral of the sum for $\ell = 0$ seems now, for large enough, positive values of s, to diverge for k = 0. If we assume that the volume Ω_3, which includes all fields is very large but finite, then the k-integration has an infrared cut-off at a very small value ε. The integral is then proportional to ε^{3-2s} or in the analytical continuation $s \to 0$, to ε^3. But this vanishes in the limiting case $\varepsilon \to 0$, i.e., in a normalization volume becoming arbitrarily large [15]. Thus we only take the first terms into account in (6.53), and calculate it just as we do the corresponding one for ζ_2^β

$$\zeta_1^\beta(s) = 8\pi\mu^{2s}\frac{\Omega_3}{(2\pi)^3}\sum_{\ell=1}^{\infty}\left\{\tfrac{2\pi}{\beta}\ell\right\}^{3-2s}\int_0^\infty dx\, x^2\left[1+x^2\right]^{-s}$$

$$= 4\pi\mu^{2s}\frac{\Omega_3}{(2\pi)^3}\left(\frac{2\pi}{\beta}\right)^{3-2s}\zeta(2s-3)\,\frac{\Gamma(\frac{3}{2})\,\Gamma(s-\frac{3}{2})}{\Gamma(s)}$$

Taking the derivative gives

$$\zeta_1^{\beta\prime}(0) = 4\pi\frac{\Omega_3}{(2\pi)^3}\left(\frac{2\pi}{\beta}\right)^3\zeta(-3)\,\Gamma(\tfrac{3}{2})\,\Gamma(-\tfrac{3}{2})\,\frac{d}{ds}\frac{1}{\Gamma(s)}\bigg|_{s=0}$$

which with the above calculated values and $\zeta(-3) = 1/120$

(from $\zeta(1-2m) = -B_{2m}/2m$, $m=2$, $B_4 = -1/30$) leads to

$$\zeta_1^{\beta'}(0) = \Omega_3 \beta^{-3} \frac{\pi^2}{3\cdot 15}$$

From

$$\mathcal{L}^{(1)}_{m=0}(B=0,T) = \Omega_3^{-1}\beta^{-1}\zeta_1^{\beta'}(0)$$

(compare Appendix), for massless, scalar QED with vanishing external field and finite temperature one then gets:

$$\mathcal{L}^{(1)}_{m=0}(B=0,T) = \frac{2}{6}\frac{1}{15}\pi^2 k^4 T^4 \tag{6.54}$$

which again coincides with the result in [19], achieved in another manner. In order to re-write (6.54), we use the integral [36]

$$\int_0^\infty \frac{x^{\nu-1}dx}{e^{\mu x}-1} = \mu^{-\nu}\Gamma(\nu)\zeta(\nu), \quad \mathrm{Re}\,\mu > 0, \; \mathrm{Re}\,\nu > 1$$

for $\mu = 1$ and $\nu = 4$; with $\Gamma(4) = 6$ and $\zeta(4) = \pi^4/90$ (for example from [36])

$$\zeta(2m) = \frac{2^{2m-1}\pi^{2m}|B_{2m}|}{(2m)!}$$

Thus the Lagrangian, i.e. the negative free energy per unit volume is

$$\mathcal{L}^{(1)}_{m=0}(B=0,T) = \frac{1}{3\pi^2}\int_0^\infty \frac{\varepsilon^3 d\varepsilon}{e^{\varepsilon/kT}-1} \tag{6.55}$$

with the correct Bose-Einstein distribution function.

We have now shown that, also when taking temperature effects into account, the Zeta function regularization represents an efficient method for evaluating determinants as they occur in calculating one-loop vacuum effects. Its particular advantage lies in the fact that, aside from calculatory simplification in comparison to known Green's functions methods, infinite counter terms must never be resorted to, as was the case for example, in our treatment in the last section.

To conclude, we should like to mention another formal aspect related to eq. (6.52) (or its analogue in spinor QED). In switching to a finite temperature, we have replaced the variable k_0 which can take on arbitrary real values by the expression $\frac{2\pi}{\beta}\ell$, which takes on only discrete values for fixed β. So we have made the time component of the momentum four vector over which we integrate, discrete and thus introduced a new parameter into the theory in the form of β. One can now ask whether an analogous substitution in a space component of the momentum vector also takes on a physical meaning. In Appendix C, we shall show that this is, in fact, so if one considers the Casimir effect [20-23] in the light of path integral quantization followed by Zeta function regularization. Thereby, the distance a between the two plates introduced into the vacuum would play an analogous role to the parameter β.

(7) Two-Loop Effective Lagrangian

Thanks to the preparations in sections (3) and (4), we can now quite simply derive a compact expression for the two-loop effective Lagrangian $L^{(2)}$ of spinor QED. The complete generating functional for the Green's functions of this theory [41] serves as starting point here, which, in the presence of an external field described by the potential A^{μ} can be written as

$$Z[j,\eta,\bar{\eta}] = N_V^{-1}\, e^{\frac{i}{2} j D_+ j}\, e^{-\frac{i}{2}\frac{\delta}{\delta J} D_+ \frac{\delta}{\delta J}} \cdot$$

$$\cdot\, e^{i\bar{\eta} G_+[J+A]\eta}\, e^{iW[J+A]}$$

$$iW[A] = -\operatorname{Tr} \ln (1 - e\not{A} G_+)^{-1} \qquad (7.1)$$

$$J \equiv D_+ j$$

One obtains arbitrary n-point functions from this by differentiation with respect to the commuting or anti-commuting (Graß-mann) c-number sources j and $\eta(\bar{\eta})$ and putting the sources equal to zero thereafter. The normalization constant N_V, which can be identified with the vacuum amplitude $\langle 0_+|0_-\rangle_A^{j,\eta,\bar{\eta}=0}$ is to be chosen in such a manner that $Z[\underline{0},\underline{0},\underline{0}] = 1$. Thus it follows from (7.1) for the vacuum amplitude in the external field (see also our discussion in section (1)):

$$N_V \equiv \langle 0_+|0_-\rangle_A = e^{-\frac{i}{2}\frac{\delta}{\delta J} D_+ \frac{\delta}{\delta J}}\, e^{iW[A+J]}\Big|_{J=\underline{0}} \qquad (7.2)$$

This formula contains radiative corrections to N_V of arbitrary order which one can calculate perturbatively, i.e, by expanding the left exponential function. If one only takes the 0. term of the series into account, one gets precisely (5.1) back! Since we consider the quantum fluctuations represented by J as small perturbations compared to the classical background field A, we now expand W[A+J] around A and terminate the Taylor series after the quadratic term

$$W[A+J] = W[A] + \left(\frac{\delta}{\delta J^{\mu}} W[A+J]\right)_{J=\underline{0}} J^{\mu} + \frac{1}{2}\left(\frac{\delta^2 W[A+J]}{\delta J^{\mu}\,\delta J^{\nu}}\right)_{J=\underline{0}} J^{\mu} J^{\nu} + \cdots$$

$$=: W^{(1)}[A] + \int d^4x\, \langle j_\mu^A(x)\rangle\, J^{\mu}(x) \qquad (7.3)$$

$$+ \frac{1}{2}\int d^4x\, d^4x'\, J^{\mu}(x)\, \Pi^A_{\mu\nu}(x,x')\, J^{\nu}(x') + \cdots$$

According to Appendix D, the required functional derivatives are

$$< j_\mu^A(x) > = \frac{\delta W[A+\mathcal{J}]}{\delta \mathcal{J}^\mu(x)}\bigg|_{\mathcal{J}=0} = ie\ tr[\gamma_\mu G_+(x,x|A)]$$

$$\Pi^A_{\mu\nu}(x,x') = \frac{\delta^2 W[A+\mathcal{J}]}{\delta \mathcal{J}^\mu(x)\delta \mathcal{J}^\nu(x')}\bigg|_{\mathcal{J}=0} = ie^2\ tr[\gamma_\mu G_+(x,x'|A)\gamma_\nu G_+(x',x|A)]$$

In the framework of this approximation, we get for the vacuum amplitude:

$$\langle 0_+|0_-\rangle_A = e^{iW^{(1)}[A]} e^{-\frac{i}{2}\frac{\delta}{\delta \mathcal{J}}D_+\frac{\delta}{\delta \mathcal{J}}} e^{\frac{i}{2}\mathcal{J}\Pi^A\mathcal{J}} e^{i< j_\mu^A> \mathcal{J}^\mu}\bigg|_{\mathcal{J}=0}$$

With the aid of the identity [41]

$$\exp\left[-\frac{i}{2}\int\frac{\delta}{\delta \mathcal{J}}A\frac{\delta}{\delta \mathcal{J}}\right]\ \exp\left[\frac{i}{2}\int jBj + i\int fj\right] =$$

$$\exp\Big[\frac{i}{2}\int j\bar{B}j + i\int j(1-BA)^{-1}f + \frac{1}{2}\mathrm{Tr}\ln(1-BA)^{-1}$$

$$+\frac{i}{2}\int fA(1-BA)^{-1}f\Big]\ , \qquad \bar{B} \equiv B(1-AB)^{-1}$$

this simplifies to

$$\langle 0_+|0_-\rangle = e^{iW^{(1)}[A]} \exp\Big[\frac{i}{2}\int\mathcal{J}\bar{\Pi}^A\mathcal{J} + i\int\mathcal{J}(1-\Pi^A D_+)< j^A>$$

$$+\frac{1}{2}\mathrm{Tr}\ln(1-\Pi^A D_+)^{-1} + \frac{i}{2}\int< j^A> D_+(1-\Pi^A D_+)^{-1}< j^A>\Big]_{\mathcal{J}=0}$$

$$= e^{iW^{(1)}[A]} e^{\frac{1}{2}\mathrm{Tr}\ln(1-D_+\Pi^A)^{-1}} e^{\frac{i}{2}\int< j^A> D_+< j^A>}\cdot$$

$$\cdot e^{\frac{i}{2}\int< j^A> D_+\Pi^A D_+< j^A>} \qquad (7.4)$$

We only want to study the simplest correction beyond $W^{(1)}$ and therefore take only into account the first terms of the logarithmic series of the second exponential function, which leads to a quadratic term in e; then the exponent $\langle j\rangle D_+ \Pi D_+ \langle j\rangle$ proportional to e^4 does not contribute anything to this approximation. Therefore

$$
\begin{aligned}
\langle 0_+|0_-\rangle_A &= \exp\left[i\int d^4x\, \mathcal{L}^{(1)}(x)\right] \\
&\cdot \exp\Big[i\int d^4x \Big\{\frac{e^2}{2}\int d^4y\, \mathrm{tr}\left[\gamma_\mu G_+(x,y|A)\gamma_\nu \right. \\
&\qquad \left.\cdot\, G_+(y,x|A)\right] D_+^{\mu\nu}(x-y)\Big\}\Big] \\
&\cdot \exp\left[\frac{i}{2}\langle j^A\rangle D_+ \langle j^A\rangle\right]
\end{aligned}
\tag{7.5}
$$

The first term added to $L^{(1)}$ can be characterized by the Feynman diagram

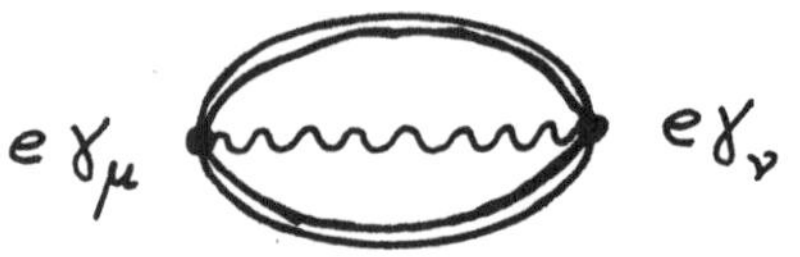

(7.6)

which suggests that the interaction of the electron with the external field is taken into account to all orders, while the interaction with the photon field is only considered up to the order e^2. Analogously, the graph

(7.7)

represents the third part of (7.5) which, however, as we now wish to show, contributes nothing for constant external fields. First, it is clear that for B (x) = const also the diagonal matrix element of the propagator (cf. (2.47)) and with it $\langle j_\mu^A\rangle$ is independent of the space-time point x. On the other hand, $\langle j_\mu^A\rangle$ is a four-vector field which, under an arbitrary Lorentz tranformation (ℓ_{μ}^{ν}) transforms according to

$$< j_{\mu}^{A\,\prime}(x') > = \ell_{\mu}{}^{\nu} < j_{\nu}^{A}(x) > \tag{7.8}$$

If the x-dependence vanishes, then

$$< j_{\mu}^{A\,\prime}(x') > = < j_{\mu}^{A}(x) > = \mathrm{const} =: < j_{\mu}^{A} >$$

which, however, implies $<j_{\mu}^{A}> = 0$ according to (7.8), qed.

From (7.5), then follows the expression for the first correction of the Lagrangian going beyond $L^{(1)}$

$$\mathcal{L}^{(2)} = \frac{e^2}{2} \int d^4x'\, \mathrm{tr}\left[\gamma_{\mu} G_{+}(x,x'|A)\, \gamma_{\nu}\, G_{+}(x',x|A)\right] D_{+}^{\mu\nu}(x-x') \tag{7.9}$$

which is in accord with the 2-loop diagram (7.6). In particular, in the Feynman gauge with

$$D_{+}^{\mu\nu} = g^{\mu\nu} D_{+}$$

this means

$$\mathcal{L}^{(2)} = \frac{e^2}{2} \int d^4x'\, \mathrm{tr}\left[\gamma_{\mu} G_{+}(x,x'|A)\, \gamma^{\mu}\, G_{+}(x',x|A)\right] D_{+}(x-x')$$

Transformation to the momentum representation by

$$G_{+}(x,x'|A) = \Phi(x,x') \int \frac{d^4p}{(2\pi)^4}\, e^{ip\cdot(x-x')}\, g(p)$$

gives, due to $\Phi(x,x')\,\Phi(x',x) = 1$, the gauge invariant expression

$$\mathcal{L}^{(2)} = \frac{e^2}{2} \int d^4x'\, \mathrm{tr}\left[\gamma_{\mu} \int \frac{d^4p}{(2\pi)^4} e^{ip(x-x')} g(p)\, \gamma^{\mu} \int \frac{d^4q}{(2\pi)^4} e^{iq(x'-x)} g(q)\right] D_{+}(x-x')$$

$$= \frac{e^2}{2}\, \mathrm{tr}\left[\gamma_{\mu} \int \frac{d^4p}{(2\pi)^4} e^{ipx} g(p)\, \gamma^{\mu} \int \frac{d^4q}{(2\pi)^4} e^{-iqx} g(q)\right] \cdot$$

$$\cdot \int d^4x'\, D_{+}(x-x')\, e^{-ix'(p-q)}$$

Now we introduce the substitution $\xi = x' - x$ into the x'-integral and use $D_+(x'-x) = D_+(x-x')$

$$\int d^4x'\, D_+(x-x')\, e^{-ix'(p-q)} = \int d^4\xi\, D_+(\xi)\, e^{-i\xi(p-q)}\, e^{-ix(p-q)}$$

$$= D_+(p-q)\, e^{-ix(p-q)}$$

Substitution results in

$$\mathcal{L}^{(2)} = \frac{e^2}{2}\, tr\left[\gamma_\mu \int \frac{d^4p}{(2\pi)^4}\, e^{ipx}\, g(p)\, \gamma^\mu \int \frac{d^4q}{(2\pi)^4}\, e^{-iqx}\, g(q)\right] \cdot$$

$$\cdot\, D_+(p-q)\, e^{-ix(p-q)}$$

$$= \frac{e^2}{2} \int \frac{d^4p}{(2\pi)^4} \int \frac{d^4q}{(2\pi)^4}\, tr\left[\gamma_\mu\, g(p)\, \gamma^\mu\, g(q)\right] D_+(p-q)$$

So $L^{(2)}$ no longer is dependent on x and is thus translational invariant, as was to be expected because of the assumed constant background field. If we now set $k := p-q$ and choose p and k as integration variables, then

$$\mathcal{L}^{(2)}(B) = \frac{e^2}{2} \int \frac{d^4p\, d^4k}{(2\pi)^8}\, tr\left[\gamma_\mu\, g(p)\, \gamma^\mu\, g(p-k)\right] D_+(k) \qquad (7.10)$$

which is in compliance with the momentum space Feynman diagram

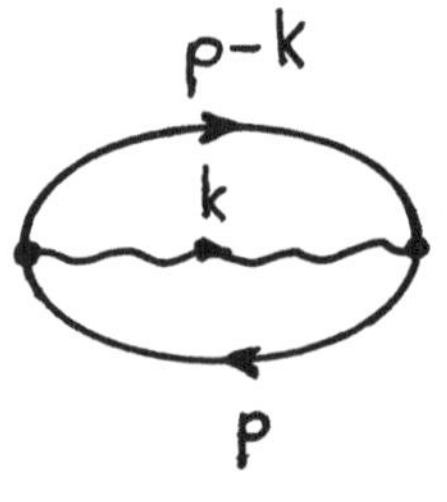

If we recall the definition of the polarization tensor

$$\Pi^{(2)}_{\mu\nu}(k) = -ie^2 \int \frac{d^4p}{(2\pi)^4}\, tr\left[\gamma_\mu\, g(p)\, \gamma_\nu\, g(p-k)\right]$$

then we can also write (7.10) in the form

$$\mathcal{L}^{(2)}(B) = \frac{i}{2}\int\frac{d^4k}{(2\pi)^4}\, D_+(k)\left\{-ie^2\int\frac{d^4p}{(2\pi)^4}\, \mathrm{tr}\left[\gamma_\mu\, g(p)\,\gamma^\mu\, g(p-k)\right]\right\}$$

$$= \frac{i}{2}\int\frac{d^4k}{(2\pi)^4}\,\frac{1}{k^2-i\varepsilon}\,\Pi^{(2)}{}_\mu{}^\mu(k) \tag{7.11}$$

We found in section (4) for the unrenormalized polarization tensor

$$\Pi^{(2)}_{\mu\nu}(k) = \frac{\alpha}{2\pi}\int_0^\infty\frac{ds}{s}\int_{-1}^{1}\frac{dv}{2}\, e^{-is\varphi_0}\left[(g_{\mu\nu}k^2 - k_\mu k_\nu)N_0 - \right.$$

$$\left. -(g^{\parallel}_{\mu\nu}k_\parallel^2 - k_{\parallel\mu}k_{\parallel\nu})N_1 + (g^{\perp}_{\mu\nu}k_\perp^2 - k_{\perp\mu}k_{\perp\nu})N_2\right] \tag{7.12}$$

with

$$\varphi_0 = m^2 + \frac{1}{4}(1-v^2)k_\parallel^2 + \frac{\cos zv - \cos z}{2z\sin z}\, k_\perp^2$$

The counter terms c.t. were left out because we have to renormalize $L^{(2)}$ again anyway. It follows from (7.12), because $g^\mu_\mu = 4$ and $g^{\perp\mu}_\mu = g^{\parallel\mu}_\mu = 2$ for the required trace:

$$\Pi^{(2)}{}_\mu{}^\mu(k) = \frac{\alpha}{2\pi}\int_0^\infty\frac{ds}{s}\int_{-1}^{1}\frac{dv}{2}\, e^{-is\varphi_0}\left[3k^2N_0 - k_\parallel^2 N_1 + k_\perp^2 N_2\right]$$

$$= \frac{\alpha}{2\pi}\int_0^\infty\frac{ds}{s}\int_{-1}^{1}\frac{dv}{2}\, e^{-is\varphi_0}\left[(3N_0 - N_1)k_\parallel^2 + (3N_0 + N_2)k_\perp^2\right]$$

Furthermore, it is useful to set

$$\varphi_0 = m^2 + a k_\parallel^2 + b\, k_\perp^2 \tag{7.13}$$

with

$$a := \frac{1}{4}(1-v^2)$$

$$b := \frac{\cos zv - \cos z}{2z\sin z} \tag{7.14}$$

If one chooses the proper time representation for the photon propagator in (7.11)

$$\frac{1}{K^2 - i\varepsilon} = i \int_0^\infty ds' \, e^{-is'(K^2 - i\varepsilon)}$$

then it follows for the Lagrangian

$$\mathcal{L}^{(2)}(B) = -\frac{\alpha}{4\pi} \int_0^\infty \frac{ds}{s} \int_{-1}^{1} \frac{dv}{2} \, e^{-im^2 s} \, I \, ,$$

$$I := \int_0^\infty ds' \int \frac{d^4K}{(2\pi)^4} \, e^{-is'(K^2 - i\varepsilon)} \, e^{-is(aK_\parallel^2 + bK_\perp^2)} \left[(3N_0 - N_1)K_\parallel^2 + (3N_0 + N_2)K_\perp^2 \right] \tag{7.15}$$

Now the k-integration can be easily performed with (3.19):

$$\int \frac{d^4K}{(2\pi)^4} \, e^{-is'K^2} \, e^{-is(aK_\parallel^2 + bK_\perp^2)} \, K_\parallel^2$$

$$= \frac{i}{s} \frac{d}{da} \int \frac{d^4K}{(2\pi)^4} \, e^{-iK_\parallel^2(s' + as) - iK_\perp^2(s' + bs)}$$

If we set $A_1 := s' + as$ and $A_2 := s' + bs$, then

$$\int \frac{d^4K}{(2\pi)^4} \, e^{-is'K^2} \, e^{-is(aK_\parallel^2 + bK_\perp^2)} \, K_\parallel^2$$

$$= \frac{i}{s} \frac{d}{da} \frac{1}{(2\pi)^4} \int_{-\infty}^{\infty} dK^0 \, e^{iA_1(K^0)^2} \int_{-\infty}^{\infty} dK^1 \, e^{-iA_1(K^1)^2} \cdot$$

$$\cdot \left(\int_{-\infty}^{\infty} dK^2 \, e^{-iA_2(K^2)^2} \right)^2 \tag{7.16}$$

$$= -\frac{1}{(4\pi)^2} \frac{1}{(s' + as)^2 (s' + bs)}$$

Analogously, we calculate

$$\int \frac{d^4k}{(2\pi)^4} \, e^{-is'k^2} \, e^{-is(aK_\parallel^2 + bK_\perp^2)} \, K_\perp^2 = -\frac{1}{(4\pi)^2} \frac{1}{(s' + as)(s' + bs)^2} \tag{7.17}$$

Substitution gives us for I

$$I = -\frac{1}{(4\pi)^2}\int_0^\infty ds' \left\{ \frac{3N_0 - N_1}{(s'+sa)^2(s'+sb)} + \frac{3N_0 + N_2}{(s'+sa)(s'+sb)^2} \right\}$$

The integration over s' is still elementary and gives [36]

$$I = -\frac{1}{(4\pi)^2}\Big[(3N_0 - N_1)\frac{1}{s^2}\left\{\frac{1}{a(b-a)} - \frac{1}{(b-a)^2}\ln\frac{b}{a}\right\}$$

$$+ (3N_0 + N_2)\frac{1}{s^2}\left\{\frac{1}{b(a-b)} + \frac{1}{(b-a)^2}\ln\frac{b}{a}\right\}\Big]$$

From (7.15) we then get for $L^{(2)}$

$$\mathcal{L}^{(2)} = \frac{\alpha}{(4\pi)^3}\int_0^\infty \frac{ds}{s^3}\int_{-1}^{1}\frac{dv}{2}\, e^{-im^2 s}\, K \tag{7.18}$$

with

$$\begin{aligned} K &= (3N_0 - N_1)\left\{\frac{1}{a(b-a)} - \frac{1}{(b-a)^2}\ln\frac{b}{a}\right\} \\ &\quad + (3N_0 + N_2)\left\{\frac{1}{b(a-b)} + \frac{1}{(b-a)^2}\ln\frac{b}{a}\right\} \\ &= K_1 + K_2 \ln\frac{b}{a} \end{aligned} \tag{7.19}$$

where

$$K_1 = (3N_0 - N_1)\frac{1}{a(b-a)} + (3N_0 + N_2)\frac{1}{b(a-b)}$$

$$K_2 = \frac{1}{(b-a)^2}(N_1 + N_2)$$

From (4.28c), we first get

$$(3N_0 - N_1) = \frac{z}{\sin z}\left[2\cos zv - 2v\cot z \sin zv + (1-v^2)\cos z\right] \tag{7.20}$$

$$(3N_0 + N_2) = \frac{z}{\sin z}\left[2\cos zv - 2v\cot z\sin zv + \frac{2(\cos zv - \cos z)}{\sin^2 z}\right]$$

$$(N_1 + N_2) = \frac{z}{\sin z}\left[\frac{2(\cos zv - \cos z)}{\sin^2 z} - (1-v^2)\cos z\right]$$

Further, the definitions (7.14) give

$$\frac{1}{(b-a)^2} = \frac{16\; z^2 \sin^2 z}{[2(\cos zv - \cos z) - (1-v^2)\, z \sin z]^2}$$

$$\frac{1}{a(b-a)} = \frac{16\; z \sin z}{(1-v^2)[2(\cos zv - \cos z) - (1-v^2)\, z \sin z]}$$

$$\frac{1}{b(a-b)} = \frac{-8 z^2 \sin z}{(\cos zv - \cos z)[2(\cos zv - \cos z) - (1-v^2)\, z \sin z]}$$

Together with the above equations, it follows from (7.20):

$$K_1 = \frac{16 z}{\sin z}\Bigg[\frac{z \sin z\,(\cos zv - v \cot z\, \sin zv)}{(1-v^2)(\cos zv - \cos z)} + \frac{z(\cos z \cdot \sin z - z)}{2(\cos zv - \cos z) - (1-v^2)\, z \sin z}\Bigg]$$

$$K_2 = \frac{16 z^3}{\sin z}\; \frac{[2(\cos zv - \cos z) - (1-v^2) \cos z\, \sin^2 z]}{[2(\cos zv - \cos z) - (1-v^2)\, z \sin z]^2} \tag{7.21}$$

$$\frac{b}{a} = \frac{2(\cos zv - \cos z)}{(1-v^2)\; z \sin z}$$

We can then write for the unrenormalized Lagrangian

$$\mathcal{L}^{(2)} = \frac{\alpha}{(4\pi)^3} \int_0^\infty \frac{ds}{s^3} \int_{-1}^{1} \frac{dv}{2}\, e^{-im^2 s} \left\{ K_1 + K_2 \ln \frac{b}{a} \right\} \tag{7.22}$$

where the s-integration for $s \to 0$ apparently diverges. In order to regularize (7.22), we introduce a finite lower limit $s_0 > 0$ for the proper time integrals of the electron propagators contained in the polarization tensor (4.5). As a result, (4.8) must be replaced by

$$\int_{s_0}^\infty ds_1 \int_{s_0}^\infty ds_2 \;\ldots\; = \int_{2s_0}^\infty s\,ds \int_{-(1-\frac{2s_0}{s})}^{+(1-\frac{2s_0}{s})} \frac{dv}{2} \;\ldots \tag{7.23}$$

which, for the Lagrangian, leads to

$$\mathcal{L}^{(2)} = \frac{\alpha}{(4\pi)^3} \int_{2s_0}^{\infty} \frac{ds}{s^3} \int_{-(1-\frac{2s_0}{s})}^{+(1-\frac{2s_0}{s})} \frac{dv}{2}\, e^{-im^2 s}\, K(z,v)$$

or

$$\mathcal{L}^{(2)} = \frac{\alpha}{(4\pi)^3} \int_{2s_0}^{\infty} \frac{ds}{s^3} \int_{0}^{1-\frac{2s_0}{s}} dv\, e^{-im^2 s}\, K(z,v) \tag{7.24}$$

Here we used the fact that K is a symmetrical function in v (compare (7.21)). To now get a convergent integral for $s \to 0$, we first subtract the term constant with respect to $z = eBs$ from $K(z,v)$; thereby, $L^{(2)}$ is changed only by a constant and now has the property $L^{(2)}(B=0) = 0$. In the next step , we also take the quadratic term in z of K out of the integral. Since K is also even in z, the series expansion of the remaining integrands thus begins with a term proportional to $\frac{1}{s^3} z^4 = (eB)^4 s$ which vanishes for $s \to 0$, i.e. leads to a convergent s-integral. The series expansion of K is performed explicitly in Appendix E and gives

$$K(z,v) = K_{02}(z,v) + O(z^4)$$

with

$$K_{02}(z,v) = \frac{48}{1-v^2} + 2z^2 \tag{7.25}$$

Now we can write (7.24) as

$$\mathcal{L}^{(2)} = \frac{\alpha}{(4\pi)^3} \int_{2s_0}^{\infty} \frac{ds}{s^3} \int_{0}^{1-2s_0/s} dv\, e^{-im^2 s} \cdot (2z^2) \tag{7.26}$$

$$+ \frac{\alpha}{(4\pi)^3} \int_{2s_0}^{\infty} \frac{ds}{s^3} \int_{0}^{1-2s_0/s} dv\, e^{-im^2 s} \left[K(z,v) - K_{02}(z,v)\right]$$

Analogously to $L^{(1)}$, we shall later get rid of the divergent integral over the separated quadratic term by means of a charge and field strength renormalization; furthermore, a mass renormalization will prove necessary. This means that we have until now not calculated with the physical parameters e, B, m, but with bare quantities e_o, B_o, m_o. We therefore write (7.26) from now on in the form

$$\mathcal{L}^{(2)} = \mathcal{F}(s_0) + \frac{\alpha_0}{(4\pi)^3} \int_{2s_0}^{\infty} \frac{ds}{s^3} \int_0^{1-2s_0/s} dv\, e^{-im_0^2 s} \left[K(z,v) - K_{02}(z,v)\right] \tag{7.27}$$

with

$$\mathcal{F}(s_0) := \frac{\alpha_0}{(4\pi)^3} \int_{2s_0}^{\infty} \frac{ds}{s^3} \int_0^{1-2s_0/s} dv\, e^{-im_0^2 s}\, 2(e_0 B_0 s)^2 \tag{7.28}$$

and $z = e_0 B_0 s$. If one considers (7.21), it becomes evident that the v-integration in (7.27) diverges for $s_o = 0$ at the upper limit (i.e., for $v \to 1$). If we calculate the Laurent series of $K = K(v^2)$, we find a non-vanishing singular part (compare Appendix E):

$$K(z,v) = \frac{24}{1-v^2}\left[\frac{z^2}{\sin^2 z} + z \cot z\right] + O((1-v^2)^0)$$

So

$$K(z,v) - K_{02}(z,v) = \frac{f(z)}{1-v^2} + O(z^4, (1-v^2)^0)$$

with

$$f(z) := 24\left[\frac{z^2}{\sin^2 z} + z\cot z - 2\right] \tag{7.29}$$

Now we regularize the v-integration by also taking the term $f(z)/(1-v^2)$ out of the integral

$$\mathcal{L}^{(2)} = F(s_0) + \frac{\alpha}{(4\pi)^3} \int_{2s_0}^{\infty} \frac{ds}{s^3} \int_0^{1-2s_0/s} dv\, e^{-im_0^2 s} \frac{f(z)}{1-v^2}$$

$$+ \frac{\alpha_0}{(4\pi)^3} \int_0^{\infty} \frac{ds}{s^3} \int_0^1 dv\, e^{-im_0^2 s} \left[K(z,v) - K_{02}(z,v) - \frac{f(z)}{1-v^2}\right] \qquad (7.30)$$

In the third term s_o could be set equal to zero, since now both integrations converge

(i) For $s \to o$, $(K-K_{o2})$ as well as f vanish proportionally to s^4

(ii) For $v \to 1$, $(K-K_{o2})$ is exactly compensated for by $f(z) / (1-v^2)$.

Now we turn our attention to the double integral over the singular part of the Laurent series, where the v-integration can easily be performed

$$G(s_0) := \frac{\alpha_0}{(4\pi)^3} \int_{2s_0}^{\infty} \frac{ds}{s^3} \int_0^{1-2s_0/s} dv\, e^{-im_0^2 s} \frac{f(z)}{1-v^2}$$

$$= \frac{\alpha_0}{(4\pi)^3} \int_{2s_0}^{\infty} \frac{ds}{s^3} f(e_0 B_0 s) e^{-im_0^2 s} \int_0^{1-2s_0/s} \frac{dv}{1-v^2}$$

Now

$$\int_0^{1-2s_0/s} \frac{dv}{1-v^2} = -\frac{1}{2}\left[\ln\left|\frac{v-1}{v+1}\right|\right]_0^{1-2s_0/s}$$

$$= -\frac{1}{2} \ln \frac{2s_0}{2s(1-\ s_0/s)}$$

$$\underset{s_0 \to 0}{=} -\frac{1}{2} \ln \frac{s_0}{s_0}$$

$$= -\frac{1}{2} \ln(i\gamma m_0^2 s_0) + \frac{1}{2} \ln(i\gamma m_0^2 s)$$

so that it follows

$$G(s_0) = -\frac{\alpha_0}{2(4\pi)^3}\, \ln(i\gamma m_0^2 s_0) \int_{2s_0}^{\infty} \frac{ds}{s^3}\, e^{-im_0^2 s} f(e_0 B_0 s)$$

$$+ \frac{\alpha_0}{2(4\pi)^3} \int_{2s_0}^{\infty} \frac{ds}{s^3}\, e^{-im_0^2 s} \ln(i\gamma m_0^2 s) f(e_0 B_0 s) \tag{7.31}$$

$$=: I_1 + I_2$$

In the second integral I_2, s_0 can be set equal to zero because the integrand vanishes like $s \ln s$ for $s \to o$. (Note $f(z) \sim z^4$ for $z \approx 0$) To rearrange I_1 we write

$$f(e_0 B_0 s) \equiv f(z) = 24\left[\frac{z^2}{\sin^2 z} + z \cot z - 2\right]$$

$$= -24\, z^3 \frac{d}{dz}\left[z^{-1} \cot z - z^{-2}\right]$$

$$= -24\, s^3 \frac{d}{ds}\left[\frac{1}{s^2}\left\{(e_0 B_0 s) \cot(e_0 B_0 s) + \frac{1}{3}(e_0 B_0 s)^2 - 1\right]\right.$$

and thus get

$$I_1 = \frac{24\alpha_0}{2(4\pi)^3} \ln(i\gamma m_0^2 s_0) \int_{2s_0}^{\infty} \frac{ds}{s^3}\, e^{-im_0^2 s}\, s^3 \frac{d}{ds}\left[\frac{1}{s^2}\left\{(e_0 B_0 s) \cdot \cot(e_0 B_0 s) + \frac{1}{3}(e_0 B_0 s)^2 - 1\right\}\right]$$

This integral converges for $s_0 = 0$ and yields after a partial integration in which the boundary terms vanish,

$$I_1 = -\frac{3\alpha_0}{16\pi^3} \ln(i\gamma m_0^2 s_0) \int_{2s_0}^{\infty} \frac{ds}{s^2} \left(\frac{d}{ds} e^{-im_0^2 s}\right) \cdot \left\{(e_0 B_0 s) \cot(e_0 B_0 s) + \frac{1}{3}(e_0 B_0 s)^2 - 1\right\}$$

$$= \left[\frac{3\alpha m_0}{4\pi} \ln \frac{1}{i\gamma m_0^2 s_0}\right] \cdot \frac{(-im_0)}{4\pi^2} \int_0^{\infty} \frac{ds}{s^2}\, e^{-im_0^2 s} .$$

$$\cdot \left\{(e_0 B_0 s)\cot(e_0 B_0 s) + \frac{1}{3}(e_0 B_0 s)^2 - 1\right\}$$

According to (5.23)

$$\mathcal{L}_R^{(1)}(m_0, e_0 B_0) = \frac{1}{8\pi^2}\int_0^\infty \frac{ds}{s^3}\, e^{-im_0^2 s}\left[(e_0 B_0 s)\cot(e_0 B_0 s) + \frac{1}{3}(e_0 B_0 s)^2 - 1\right]$$

and according to (3.48a) it holds for the mass displacement that

$$\delta m(m_0, \alpha_0, s_0) = \frac{3\alpha_0 m_0}{4\pi}\left[\ln\frac{1}{i\gamma m_0^2 s_0} + \frac{5}{6}\right] + O(\alpha_0^2)$$

where s_0 has the same significance as in (7.23), i.e. represents the lower limit of the electron propagator proper-time integration. Hence we obtain for I_1

$$\begin{aligned} I_1 &= \left[\frac{3\alpha_0 m_0}{4\pi}\ln\frac{1}{i\gamma m_0^2 s_0}\right]\frac{\partial \mathcal{L}_R^{(1)}}{\partial m_0}(m_0, e_0 B_0) \\ &= \frac{3\alpha_0 m_0}{4\pi}\left(\ln\frac{1}{i\gamma m_0^2 s_0} + \frac{5}{6}\right)\frac{\partial \mathcal{L}_R^{(1)}}{\partial m_0}(m_0, e_0 B_0) \\ &\quad - \frac{5}{6}\frac{3\alpha_0 m_0}{4\pi}\frac{\partial \mathcal{L}_R^{(1)}}{\partial m_0}(m_0, e_0 B_0) \\ &= \delta m(m_0, \alpha_0, s_0)\frac{\partial \mathcal{L}_R^{(1)}}{\partial m_0}(m_0, e_0 B_0) \\ &\quad - \frac{5}{6}\frac{3\alpha_0 m_0}{4\pi}\frac{\partial \mathcal{L}_R^{(1)}}{\partial m_0}(m_0, e_0 B_0) \end{aligned}$$

For (7.31) one can then write

$$\begin{aligned} G(s_0) &= \delta m(m_0, s_0, \alpha_0)\frac{\partial \mathcal{L}_R^{(1)}}{\partial m_0}(m_0, e_0 B_0) \\ &\quad - \frac{5}{6}\frac{3\alpha_0 m_0}{4\pi}\frac{\partial \mathcal{L}_R^{(1)}}{\partial m_0}(m_0, e_0 B_0) \\ &\quad + \frac{3\alpha_0}{16\pi^3}\int_0^\infty \frac{ds}{s^3}\, e^{-im_0^2 s}\ln(i\gamma m_0^2 s)\left[\frac{z^2}{\sin^2 z} + z\cot z - 2\right] \end{aligned}$$

with $z = e_0 B_0 s$.

Because of (7.30), we have

$$\mathcal{L}^{(2)} = F(s_0) + G(s_0) + \frac{\alpha_0}{(4\pi)^3}\int_0^\infty \frac{ds}{s^3}\int_0^1 dv\, e^{-im_0^2 s}\left[K(z,v) - K_{02}(z,v) - \frac{f(z)}{1-v^2}\right].$$

For the whole two-loop effective (i.e. calculated to the first order in the coupling constant α between the fermion and radiation field) Lagrangian, one gets, keeping (5.19) in mind

$$\begin{aligned}\mathcal{L} &= \mathcal{L}^{(0)} + \mathcal{L}^{(1)} + \mathcal{L}^{(2)} + O(\alpha^2)\\
&= -\tfrac{1}{2}B_0^2\\
&\quad -\frac{1}{8\pi^2}\int_{s_0}^\infty \frac{ds}{s^3}\,\frac{1}{3}(e_0B_0 s)^2 e^{-im_0^2 s} + \mathcal{L}_R^{(1)}(m_0, e_0B_0)\\
&\quad +\frac{\alpha_0}{(4\pi)^3}\int_{2s_0}^\infty \frac{ds}{s^3}\int_0^{1-2s_0/s} dv\, e^{-im_0^2 s}\, 2(e_0B_0 s)^2 + \delta m(m_0,\alpha_0,s_0)\frac{\partial \mathcal{L}_R^{(1)}}{\partial m_0}(m_0, e_0B_0)\\
&\quad -\frac{5}{6}\,\frac{3\alpha_0 m_0}{4\pi}\,\frac{\partial \mathcal{L}_R^{(1)}}{\partial m_0}(m_0,e_0B_0) + \frac{3\alpha_0}{16\pi^3}\int_0^\infty \frac{ds}{s^3} e^{-im_0^2 s}\ln(i\gamma m_0^2 s)\cdot\\
&\quad \cdot\left[\frac{z^2}{\sin^2 z} + z\cot z - 2\right]\\
&\quad +\frac{\alpha_0}{(4\pi)^3}\int_0^\infty \frac{ds}{s^3}\int_0^1 dv\, e^{-im_0^2 s}\left[K(z,v) - K_{02}(z,v) - \frac{f(z)}{1-v^2}\right] + O(\alpha^2)\end{aligned}$$

If we use

$$\mathcal{L}_R^{(1)}(m_0) + \delta m\,\frac{\partial \mathcal{L}_R^{(1)}}{\partial m_0}(m_0) = \mathcal{L}_R^{(1)}(m_0+\delta m) + O(\alpha^2)$$

it follows that

$$\begin{aligned}\mathcal{L} &= -\tfrac{1}{2}B_0^2\left\{1 + \frac{e_0^2}{12\pi^2}\int_{s_0}^\infty \frac{ds}{s} e^{-im_0^2 s} - \frac{4e_0^2\alpha_0}{(4\pi)^3}\int_{2s_0}^\infty \frac{ds}{s}\int_0^{1-2s_0/s} dv\, e^{-im_0^2 s}\right\}\\
&\quad + \mathcal{L}_R^{(1)}(m_0+\delta m, e_0B_0) - \frac{5}{3}\,\frac{3\alpha_0 m_0}{4\pi}\,\frac{\partial \mathcal{L}_R^{(1)}}{\partial m_0}(m_0, e_0B_0) +\end{aligned}$$

$$+ \frac{3\alpha_0}{16\pi^3} \int_0^\infty \frac{ds}{s^3} e^{-im_0^2 s} \ln(i\gamma m_0^2 s) \left[\frac{z^2}{\sin^2 z} + z\cot z - 2\right]$$

$$+ \frac{\alpha_0}{(4\pi)^3} \int_0^\infty \frac{ds}{s^3} \int_0^1 dv\, e^{-im_0^2 s} \left[K(z,v) - K_{0z}(z,v) - \frac{f(z)}{1-v^2}\right] + O(\alpha^2)$$

with $z = e_0 B_0 s$. Now we perform the following renormalization:

(i) The renormalized (physical) mass is defined by

$$m = m_0 + \delta m \tag{7.33}$$

(ii) The renormalized charge and field strength is introduced by

$$e = e_0\, Z_3^{+\frac{1}{2}}$$
$$B = B_0\, Z_3^{-\frac{1}{2}} \tag{7.34}$$

wich allows for $eB = e_0 B_0$.

Here

$$Z_3^{-1}(\alpha_0, m_0) = 1 + \frac{e_0^2}{12\pi^2} \int_{s_0}^\infty \frac{ds}{s} e^{-im_0^2 s} - \frac{4e_0^2\alpha_0}{(4\pi)^3} \int_{2s_0}^\infty \frac{ds}{s} e^{-im_0^2 s} \int_0^{1-2s_0/s} dv + O(\alpha_0^2)$$

$$= 1 - \frac{\alpha_0}{3\pi} Ei(-im_0^2 s_0) + \frac{\alpha_0^2}{4\pi^2} Ei(-2im_0^2 s_0) + O(\alpha_0^2)$$

For $s_0 \to 0$, the exponential integrals reduce to a logarithm [36]:

$$Z_3^{-1}(\alpha_0, m_0) = 1 + \frac{\alpha_0}{3\pi} \ln \frac{1}{i\gamma m_0^2 s_0} - \frac{\alpha_0^2}{4\pi^2} \ln \frac{1}{2i\gamma m_0^2 s_0} + O(\alpha_0^2) \tag{7.35}$$

With (7.33) and (7.34) it follows from (7.32)

$$\mathcal{L}(B) = \mathcal{L}_R^{(0)}(B) + \mathcal{L}_R^{(1)}(m, B) + \mathcal{L}_R^{(2)}(m, B) + O(\alpha^2) \tag{7.36}$$

with

$$\mathcal{L}_R^{(2)}(m, B) = \frac{\alpha}{(4\pi)^3} \int_0^\infty \frac{ds}{s^3} \int_0^1 dv\, e^{-im^2 s} \left[K(z,v) - K_{02}(z,v) - \frac{f(z)}{1-v^2}\right]$$
$$- \frac{5}{6} \frac{3\alpha m}{4\pi} \frac{(-im)}{4\pi^2} \int_0^\infty \frac{ds}{s^2} e^{-im^2 s} \left[(eBs)\cot(eBs) + \frac{1}{3}(eBs)^2 - 1\right] \tag{7.37}$$
$$+ \frac{3\alpha}{16\pi^3} \int_0^\infty \frac{ds}{s^3} e^{-im^2 s} \ln(i\gamma m^2 s) \left[\frac{(eBs)^2}{\sin^2(eBs)} + (eBs)\cot(eBs) - 2\right]$$

and z = eBs. Here, the derivative of $L_R^{(1)}$ was explicitly inserted. Furthermore, it should be noted that because of $\delta m \sim \alpha$ for each function $g \sim \alpha^0$, $\alpha g(m_0) = \alpha g(m) + O(\alpha^2)$ is valid and that trivially $\alpha_0 = \alpha + O(\alpha^2)$; this means that one may replace the factors α_0 before the integrals for $L_R^{(2)}$ by α and the mass m_0 by m.

With (7.37) then, we have been successful in deriving a relatively lucid and simple expression for the effective Lagrangian to the first order in the coupling constant α between electron and photon field (i.e., according to (7.6) in two-loop approximation).This expression is equivalent to Ritus' [4], but was arrived at in a completely different manner.

In order to further evaluate (7.37), we limit ourselves to the limiting case of strong fields, i.e. $eB/m^2 \gg 1$. Since in all the above formulae, m^2 is to be understood as an abbreviation for $m^2 - i\varepsilon$, we now rotate the integration paths for the electron proper-times s_1 and s_2 by the substitution $s_{1,2} \to -is_{1,2}$, which according to (4.6) leads to $s \to -is$ and $v \to v$; accordingly we must replace

$$z \longrightarrow -iz$$
$$\cos z \longrightarrow \cosh z$$
$$\sin z \longrightarrow -i \sinh z$$
$$\cot z \longrightarrow i \coth z$$

and get

$$K_1(-iz, v) = \frac{-16z}{\sinh z}\left[\frac{z \sinh z\,(\cosh zv - v\coth z \cdot \sinh zv)}{(1-v^2)(\cosh zv - \cosh z)} + \frac{z(\cosh z \cdot \sinh z - z)}{2(\cosh zv - \cosh z) + (1-v^2) z \sinh z}\right]$$

$$K_2(-iz, v) = \frac{-16z^3}{\sinh z}\,\frac{[2(\cosh zv - \cosh z) + (1-v^2)\cosh z \cdot \sinh^2 z]}{[2(\cosh zv - \cosh z) + (1-v^2) z \sinh z]^2}$$

$$\left(\frac{b}{a}\right)(-iz, v) = \frac{-2(\cosh zv - \cosh z)}{(1-v^2)\, z \cdot \sinh z}$$

$$K_{02}(-iz, v) = \frac{48}{1-v^2} - 2z^2 \qquad (7.38)$$

It follows, for the Lagrangian

$$\mathcal{L}_R^{(2)}(B) = -\frac{\alpha}{(4\pi)^3}\int_0^\infty \frac{ds}{s^3}\int_0^1 dv\, e^{-m^2 s}\left[K(-iz, v) - K_{02}(-iz, v) - \frac{24}{1-v^2}\left\{\frac{z^2}{\sinh^2 z} + z\coth z - 2\right\}\right]$$

$$-\frac{5}{6}\,\frac{3\alpha m}{4\pi}\,\frac{m}{4\pi^2}\int_0^\infty \frac{ds}{s^2}\, e^{-m^2 s}\left[(eBs)\coth(eBs) - \tfrac{1}{3}(eBs)^2 - 1\right]$$

$$-\frac{3\alpha}{16\pi^3}\int_0^\infty \frac{ds}{s^3}\, e^{-m^2 s}\,\ln(\gamma m^2 s)\left[\frac{z^2}{\sinh^2 z} + z\coth z - 2\right] \qquad (7.39)$$

We can also use z = eBs instead of s as integration variable

$$\mathcal{L}_R^{(2)}(B) = -\frac{\alpha}{(4\pi)^3}(eB)^2\int_0^\infty \frac{dz}{z^3}\int_0^1 dv\, e^{-\frac{m^2}{eB}z}\left[K(-iz,v) - \frac{48}{1-v^2} + \underline{2z^2}_{A}\right.$$
$$\left. - \frac{24}{1-v^2}\left\{\frac{z^2}{\sinh^2 z} + z\coth z - 2\right\}\right]$$
$$-\frac{5}{6}\frac{3\alpha m}{4\pi}\frac{m}{4\pi^2}(eB)\int_0^\infty \frac{dz}{z^2}\, e^{-\frac{m^2}{eB}z}\left[z\coth z - \frac{1}{3}z^2 - 1\right] \qquad (7.40)$$
$$\underline{-\frac{3\alpha}{16\pi^3}(eB)^2\int_0^\infty \frac{dz}{z^3} e^{-\frac{m^2}{eB}z}\ln\left(\gamma\frac{m^2}{eB}z\right)\left[\frac{z^2}{\sinh^2 z} + z\coth z - 2\right]}_{C}$$

If we are interested in an evaluation of this still exact equation only to the order $B^2\ln(B)$ then only the two underlined terms A and C make a contribution since the exponential with $eB \ll m^2$ of the integrand is small, and thus its behaviour for large values of z becomes important. If we introduce a lower integration limit $z_o > o$ and consider at the end the limes $z_o \to o$ then we get, for the first contribution:

$$A = -\frac{2\alpha}{(4\pi)^3}(eB)^2\int_0^1 dv\int_{z_0}^\infty \frac{dz}{z}\, e^{-\frac{m^2}{eB}z}$$
$$= \frac{2\alpha}{(4\pi)^3}(eB)^2\, Ei\left(-\frac{m^2}{eB}z_0\right)$$
$$= -\frac{\alpha}{32\pi^3}(eB)^2\ln\frac{eB}{m^2} + O(B^2) \qquad (7.41)$$

The second contribution is

$$C = -\frac{3\alpha}{16\pi^3}(eB)^2\left[\ln\frac{m^2}{eB}\int_0^\infty \frac{dz}{z^3} e^{-\frac{m^2}{eB}z}\left\{\frac{z^2}{\sinh^2 z} + z\coth z - 2\right\}\right.$$
$$\left. + \int_0^\infty \frac{dz}{z^3} e^{-\frac{m^2}{eB}z}\ln(\gamma z)\left\{\frac{z^2}{\sinh^2 z} + z\coth z - 2\right\}\right]$$

Since both integrals converge even without the exponential factor, we can ignore these in the case of strong fields and then have

$$\mathcal{C} = \frac{3\alpha}{16\pi^3}(eB)^2\, I \,\ln\frac{eB}{m^2} + O(B^2)$$

where

$$I = \int_0^\infty \frac{dz}{z^3}\left\{\frac{z^2}{\sinh^2 z} + z\coth z - 2\right\}$$

$$= \int_0^\infty dz\left\{\frac{1}{z\sinh^2 z} + \frac{\coth z}{z^2} - \frac{2}{z^3}\right\}$$

Partial integration in the second term gives

$$\int_{z_0}^\infty dz\,\frac{\coth z}{z^2} = \left[-\frac{1}{z}\coth z\right]_{z_0}^\infty + \int_{z_0}^\infty dz\,\frac{(-1)}{z\sinh^2 z}$$

and thus

$$I = \left[-\frac{1}{z}\coth z\right]_{z_0}^\infty - 2\int_{z_0}^\infty \frac{dz}{z^3}$$

$$= \frac{1}{3} + O(z_0)$$

which finally leads to

$$\mathcal{C} = \frac{\alpha}{16\pi^3}(eB)^2\ln\frac{eB}{m^2} + O(B^2)$$

$$= 2\frac{\alpha}{32\pi^2}(eB)^2\ln\frac{eB}{m^2} + O(B^2) \tag{7.42}$$

This all results in the following asymptotic form of the two-loop effective Lagrangian

$$\mathcal{L}_R^{(2)}(B) \approx A + \mathcal{C}$$

$$\approx \frac{\alpha^2 B^2}{8\pi^2}\ln\frac{eB}{m^2} \tag{7.43}$$

This result is identical to the one Ritus [4] gets from his calculation.

It is interesting that (7.43) has the same dependence on B as the asymptotic form (5.29) of $L_R^{(1)}$ for large values of B; retaining only terms of order $B^2 \ln(B)$, the ratio $L_R^{(2)}/L_R^{(1)}$ is B-independent for $eB \gg m^2$:

$$\frac{\mathcal{L}^{(2)}(B)}{\mathcal{L}^{(1)}(B)} \approx \frac{3}{4\pi}\,\alpha \tag{7.44}$$

As was to be expected, the two-loop effective Lagrangian, which contains an additional factor of α due to the photon coupling, is about two orders of magnitude smaller than the Euler-Heisenberg (one-loop) Lagrangian $L_R^{(1)}$. So, bearing in mind the smallness of the effects associated with $L_R^{(1)}$, it appears extremely unlikely that effects due to radiative corrections of the one-loop calculations can be detected experimentally in the near future; whether there are astrophysical objects with magnetic fields large enough for radiative corrections to be measurable, further development must show. Thus, the refinement of one-loop calculations being of minor importance, the significance of $L_R^{(2)}$ is of a more conceptional nature.

Another point should be mentioned in this context. If one wants to do a realistic calculation of $\langle 0_+|0_-\rangle_A$ including corrections due to the fluctuations of the photon field, one should also take into account particles other than electrons and positrons; admittedly, they are the lightest charged particles known and hence can be created from the vacuum with the least effort, but for muons, say, the creation rate calculated from $L_R^{(1)}$ for a given value of E is of about the same order of magnitude as the radiatively corrected creation rate for electrons. (It is only for very strong fields that the mass of the fermion becomes unimportant, cf. eq. (5.41)). It thus appears that in principle one should include all the other charged particles occuring in nature into one's calculation. Neglecting a particle of mass M is a valid approximation only as long as $eE \gg M^2$. The inclusion of other leptons would be quite straightforward; however, owing to the strong interactions, the analogous computations for hadrons would be a formidable task.

Now, instead of dwelling further upon the explicit form (7.37), let us discuss some of the salient features of our two-loop calculation. First of all, because QED is a renormalizable theory, it is clear from the very beginning that by renormalizing B, e and the fermion mass, it must be possible to get a finite effective action $W^{(2)}$ or Lagrangian $L^{(2)}$. What is less clear is what a mass renormalization looks like at the level of an effective Lagrangian. Diverging terms arising in the calculation of $L^{(2)}$ with a field dependence proportional to B^2 simply can be absorbed in $L^{(0)}$ to produce a renormalized classical Lagrangian $L_R^{(0)}$; this was already done when calculating $L^{(1)}$. However, already in eq. (7.30) it turns out that there are diverging terms with a complicated field dependence (through the function f(z) given by (7.29)) which can not be incorporated in the classical Maxwell Lagrangian. Because the only further parameter available for a redefinition is m, it should be possible to separate off a factor of $\delta m(s_0)$ from these terms. As we saw, this is indeed the case and now the power of renormalizability forces the coefficient of $\delta m(s_o)$ to be exactly $\partial L^{(1)}/\partial m$ (see the equations prior to (7.33)). Because this is the only funtional form which allows us to absorb the divergent terms into a structure already present in L_{eff}, namely $L^{(1)}(m_o)$. We think that this is an impressive example of how in a non-trivial case the fact that a theory is renormalizable constrains the appearance of divergencies.

Another interesting point is that it was also possible to construct $L_R^{(2)}$ in a gauge invariant way. The explicit dependence on the vector potential A_μ vanishes in our calculation at the point where the Fourier transform of $G_+(x,x' \mid A)$, which contains the gauge dependent factor $\phi(x,x')$ (cf. eq. (2.47)), is inserted in (7.9). Because of $\phi(x,x')\ \phi(x',x) = 1$, which immediately follows from (2.16), the only remaining field dependence of $L^{(2)}$ resides in the trivially gauge invariant factor $z \equiv eBs$. Strictly speaking, this only means that the unrenormalized $L^{(2)}$ is gauge invariant; our calculation shows, however, that renormalization can be done in a way which conserves this property. According to section (1), this implies that the renormalized expectation value of the induced vacuum current is still conserved and that the generalized Maxwell equations (1.6) are consistent. (This remark might appear trivial for the case of QED; nevertheless, there are spinor field theories (containing γ_5-couplings) where there is already at the one loop level no regularizing scheme which preserves the conservation property of $\langle j^\mu \rangle$, i.e., the gauge invariance of W_{eff}. These phenomena are known as chiral anomalies. For an introduction within the above context, see Jackiw [56].)

It is necessary to emphasize that the relatively compact representation (7.37) for $L_R^{(2)}$ (the corresponding expressions in [4] look much more complicated) could only be obtained because of the decomposition (1.57) of the two-loop diagram together with Tsai's representation for the polarization tensor. One also could think of decomposing the two-loop diagram into an electron mass operator calculated to first order in α, as computed in section (3), and an electron propagator, both to all orders in the coupling to the external field, of course. Symbolically:

It turns out, however, that this choice would be much less favourable than our decomposition (1.57); $L^{(2)}$ expressed in this way would contain a two-fold parameter integral coming from Σ and one further coming from the electron propagator. None of these can be done in an easy analytical way, whereas in our approach, one of the three (the s'-integration prior to eq. (7.18)) could be calculated in terms of elementary functions.

As a by-product of the renormalization of $L^{(2)}$, we got an explicit expression for the renormalization constant Z_3 as a function of the proper time cut-off s_0 (see (7.35)). In the next section, this will be used to calculate the QED β-function up to order α^2. This is usually done using the photon polarization tensor; however, as will be seen, there is a close analogy between electrodynamics at short distances and electrodynamics with strong fields [4]. Thus it is not surprising at all that L_{eff} contains the same information as the polarization function of the corresponding order.

In concluding let us recall that the calculations in this section can also be interpreted in the framework of Schwinger's Source Theory; here, the determination of suitable contact terms takes the place of the renormalization. We shall go into the details of these methods in Appendix F.

(8) Renormalization Group Equations

In this section we shall further improve our preceding one- and two-loop calculations for the effective Lagrangian of QED with the aid of a perturbation series for the renormalization group equation in the asymptotic region, i.e., for $eB \gg m^2$ [4,31]. Here it will be shown, for example, that in this field strength region, pair production propability undergoes no radiative corrections in a pure electric field in one-loop approximation. Just as this also applies for the n-point functions of the theory, so there are also various renormalization group equations for L for various renormalization schemes [32]: a Callan-Symanzik equation for 'On-Shell' renormalization, a 't Hooft-Weinberg equation for dimensional regularization, etc. For the actual perturbation calculation, we shall use here the Callan-Symanzik equation, since we have performed all preceding renormalizations according to the mass shell scheme. First, however, we shall describe the relation between both types of equations and demonstrate in particular that the corresponding β-functions coincide to the order α^2.

It was shown by 't Hooft with the help of dimensional regularization [12,13] that for every renormalizable field theory, a renormalization scheme with the following properties exists:

(i) The bare quantities (besides the mass) can be chosen independent of the renormalized mass.

(ii) The bare mass is proportional to the renormalized mass. ('multiplicative mass renormalization').

This method is described as renormalization by minimal subtraction (MS). Furthermore, it is important that in the framework of dimensional regularization, an initially arbitrary mass parameter μ is introduced in order to assure that, for an arbitrary dimension n in space, the renormalized parameters of the theory have the same dimension as the bare ones for n=4. This together with (i) and (ii), leads to the following Laurent series for α_o and m_o as a function of the renormalized quantities α_R and m_R:

$$\alpha_o = \mu^{4-n}\, \alpha_R \left[1 + \sum_{\ell=1}^{\infty} \frac{a_\ell(\alpha_R)}{(n-4)^\ell}\right] \tag{8.1}$$

$$m_o = m_R \left[1 + \sum_{\ell=1}^{\infty} \frac{b_\ell(\alpha_R)}{(n-4)^\ell}\right] \tag{8.2}$$

Here, α_R is dimensionless for every n (α_o has dimension $(\text{mass})^{4-n}$!) and the coefficients a_ℓ and b_ℓ are independent of m_R.

If we had used the 'On-Shell' scheme, then the renormalized quantities α_R and m_R would have been identical to the physical quantities α and m, which is now not the case because of the arbitrariness of μ i.a.. For a particular value of μ which must satisfy an equation of the form $\mu \doteq \kappa(\alpha)m$ for dimensional reasons, the renormalized mass or charge will be equal to the physical. Accordingly, there is a function $\tilde{\kappa}$ with $m_R = \tilde{\kappa}(\alpha)m$, so that it follows that

$$\alpha_0 = (\kappa(\alpha)m)^{4-n}\,\alpha\left[1+\sum_{\ell=1}^{\infty}\frac{a_\ell(\alpha)}{(n-4)^\ell}\right] \tag{8.3}$$

$$m_0 = \tilde{\kappa}(\alpha)\,m\left[1+\sum_{\ell=1}^{\infty}\frac{b_\ell(\alpha)}{(n-4)^\ell}\right]$$

The observable quantities are defined here by

$$\alpha = \frac{\alpha_0\,\mu^{n-4}}{1+\Pi(q^2=0)}$$

eq. on page 39; unlike m_R and e_R, they are independent of μ. Kaminski [31] used this dimensional scheme to calculate the effective Lagrangian in QED; the result is an expression which is dependent on μ

$$\mathcal{L}_R(e_R B_R, \alpha_R, m_R, \mu)$$

which coincides with the corresponding result of the mass shell renormalization only for a special value $\mu = \kappa(\alpha)m$

$$\mathcal{L}_{\text{on shell}}(eB, \alpha, m) = \mathcal{L}_R(e_R B_R, \alpha_R(\mu,\alpha,m), m_R(\mu,\alpha,m),\mu)\Big|_{\mu=\kappa(\alpha)m}$$

The index 'R' in 'L_R' has a different meaning here than in the earlier sections; it suggests that the renormalized Lagrangian as well as m_R and e_R depends on the choice of μ. The renormalization condition for L now reads

$$\mathcal{L}_0(e_0 B_0, \alpha_0, m_0) = \mathcal{L}_R(e_R B_R, \alpha_R, m_R, \mu) \tag{8.4}$$

The 'bare Lagrangian' L_0 and its arguments are not dependent on μ. Therefore, the right side of (8.4) must also be independent of μ

$$\mu \frac{d}{d\mu} \mathcal{L}_R (e_R B_R, \alpha_R, m_R, \mu) = 0$$

Because $e_R B_R = e_o B_o = eB$, this can also be written as

$$\mu \frac{d}{d\mu} \mathcal{L}_R (eB, \alpha_R, m_R, \mu) = 0 \tag{8.5}$$

The implicit dependencies of μ according to the chain rule lead to

$$\left[\mu \frac{\partial}{\partial\mu} + \mu \frac{\partial \alpha_R}{\partial \mu} \frac{\partial}{\partial \alpha_R} + \mu \frac{\partial m_R}{\partial \mu} \frac{\partial}{\partial m_R}\right] \mathcal{L}_R(eB, \alpha_R, m_R, \mu) = 0$$

or to

$$\left[\mu \frac{\partial}{\partial\mu} + \alpha_R \beta_{tW}(\alpha_R) \frac{\partial}{\partial \alpha_R} - \gamma_m(\alpha_R) m_R \frac{\partial}{\partial m_R}\right] \mathcal{L}_R(eB, \alpha_R, m_R, \mu) = 0 \tag{8.6}$$

with the t'Hooft Weinberg β-function

$$\beta_{tW}(\alpha_R) := \frac{\mu}{\alpha_R} \frac{\partial \alpha_R}{\partial \mu} \tag{8.7}$$

and

$$\gamma_m(\alpha_R) := -\frac{\mu}{m_R} \frac{\partial m_R}{\partial \mu} \tag{8.8}$$

Now we separate the Maxwell-Lagrangian from L_R

$$\begin{aligned} \mathcal{L}_R(eB, \alpha_R, m_R, \mu) &=: -\tfrac{1}{2} B^2 \, L_R(eB, \alpha_R, m_R, \mu) \\ &= -\frac{(eB)^2}{8\pi\alpha_R} L_R(eB, \alpha_R, m_R, \mu) \end{aligned} \tag{8.9}$$

and introduce new, dimensionless variables

$$t := (eB)^{1/2}/\mu \quad , \quad \delta := m_R/\mu \tag{8.10}$$

So

$$\mu\frac{\partial}{\partial\mu} = \mu\frac{\partial t}{\partial\mu}\frac{\partial}{\partial t} + \mu\frac{\partial\rho}{\partial\mu}\frac{\partial}{\partial\rho}$$

with

$$\mu\frac{\partial t}{\partial\mu} = -\frac{(eB)^{\frac{1}{2}}}{\mu} = -t$$

$$\mu\frac{\partial\rho}{\partial\mu} = -\frac{m_R}{\mu} = -\rho$$

The renormalization group equation for $L(t,\alpha_R,\rho) := L_R(eB,\alpha_R,m_R,\mu)$ thus reads

$$\left[-t\frac{\partial}{\partial t} - \rho\frac{\partial}{\partial\rho} + \beta_{tW}(\alpha_R)\alpha_R\frac{\partial}{\partial\alpha_R} - \gamma_m(\alpha_R)\rho\frac{\partial}{\partial\rho}\right]\frac{(eB)^2}{8\pi\alpha_R}L(t,\alpha_R,\rho) = 0$$

or

$$\left[t\frac{\partial}{\partial t} - \beta_{tW}(\alpha_R)\alpha_R\frac{\partial}{\partial\alpha_R} + [1+\gamma_m(\alpha_R)]\rho\frac{\partial}{\partial\rho} + \beta_{tW}(\alpha_R)\right]L(t,\alpha_R,\rho) = 0 \qquad (8.11)$$

This is the most general form of the 't Hooft-Weinberg equation for the effective Lagrangian valid for arbitrary values of B. If we limit ourselves to the limiting case $eB \gg m^2$, then one can assume that L is independent of m_R and so, of ρ as well. This assumption is plausible since, as we shall see, a certain analogy exists between the electrodynamics of strong fields and the electrodynamics of short distances, i.e. of large momenta. At arbitrarily high momenta, however, it is to be expected that each fixed, finite mass parameter becomes meaningless. (Analogous arguments are often used to establish the independence of the coefficients a_ℓ and b_ℓ from m_R, compare [12]). Eq. (8.11) then becomes

$$[t\frac{\partial}{\partial t}-\beta_{tW}(\alpha_R)(\alpha_R\frac{\partial}{\partial\alpha_R}-1)]L(t=\frac{(eB)^{\frac{1}{2}}}{\mu},\alpha_R)=0$$

We now substitute $\mu = \kappa(\alpha)m$ into L

$$[t\frac{\partial}{\partial t}-\beta_{tW}(\alpha)(\alpha\frac{\partial}{\partial\alpha}-1)]L(t=\frac{(eB)^{\frac{1}{2}}}{\kappa m},\alpha)=0$$

which with

$$t\frac{\partial}{\partial t} = t\frac{dm^2}{dt}\frac{\partial}{\partial m^2}$$

$$= t\frac{d}{dt}\left(\frac{eB}{\kappa^2 t^2}\right)\frac{\partial}{\partial m^2}$$

$$= (-2)m^2\frac{\partial}{\partial m^2}$$

leads to the following renormalization group equation for strong fields

$$[m^2\frac{\partial}{\partial m^2}+\frac{1}{2}\beta_{tW}(\alpha)(\alpha\frac{\partial}{\partial\alpha}-1)]\tilde{L}(\frac{eB}{\kappa^2(\alpha)\cdot m^2},\alpha)=0 \qquad (8.12)$$

Here, the function [32] β_{tW} which is independent of μ is defined by (8.7) or because $\alpha_R = Z_3\alpha_o$ (for n=4) by

$$\beta_{tW}(\alpha)=\frac{\mu}{Z_3(\alpha)}\frac{\partial Z_3(\alpha)}{\partial\mu} \qquad (8.13)$$

Eq. (8.12) is very similar to Ritus' [4] (homogenous) Callan-Symanzik equation for strong fields

$$[m^2\frac{\partial}{\partial m^2}+\frac{1}{2}\beta_{CS}(\alpha)\alpha\frac{\partial}{\partial\alpha}]\frac{1}{\alpha}\ell_{R\infty}(\frac{eB}{m^2},\alpha)=0$$

or

$$[m^2 \frac{\partial}{\partial m^2} + \frac{1}{2}\beta_{cs}(\alpha)(\alpha\frac{\partial}{\partial\alpha} - 1)]\, \ell_{R\infty}(\frac{eB}{m^2}, \alpha) = 0 \qquad (8.14)$$

Here,

$$\ell_{R\infty} = \lim_{\frac{eB}{m^2}\to\infty} \frac{\mathcal{L}_R}{\mathcal{L}_R^{(0)}}$$

where the index 'R' again has the same meaning as in the 5th and 7th sections. However, the β-function is now defined by

$$\beta_{cs}(\alpha) = \frac{m}{Z_3(\alpha)} \frac{\partial Z_3(\alpha)}{\partial m} = 2\, \frac{m^2}{Z_3(\alpha)} \frac{\partial Z_3(\alpha)}{\partial m^2} \qquad (8.15)$$

This function was calculated by de Rafael and Rosner [42] with the help of the polarization tensor to the order α^3; the result is

$$\beta_{cs}(\alpha) = \frac{2}{3}(\frac{\alpha}{\pi}) + \frac{1}{2}(\frac{\alpha}{\pi})^2 - \frac{121}{144}(\frac{\alpha}{\pi})^3 + O(\alpha^4) \qquad (8.16)$$

It is now interesting to note that β_{cs} coincides with β_{tw} to the order α^2; to prove this we calculate

$$\frac{\beta_{cs}(\alpha)}{\beta_{tw}(\alpha)} = \frac{m}{Z_3}\frac{\partial Z_3}{\partial m}\left[\frac{\mu}{Z_3}\frac{\partial Z_3}{\partial \mu}\right]^{-1}$$

$$= \frac{m}{\mu}\frac{\partial \mu}{\partial m}$$

With $\mu = \kappa(\alpha=\alpha_R(\mu=\kappa m))m$ it now follows that

$$\frac{\beta_{cs}(\alpha)}{\beta_{tw}(\alpha)} = \frac{1}{\kappa}\left[\kappa + m\frac{\partial\kappa}{\partial m}\right]$$

$$= 1 + \frac{m}{\kappa}\frac{\partial\alpha_R}{\partial m}\frac{\partial\kappa}{\partial\alpha_R}$$

$$= 1 + \frac{\kappa'}{\kappa}\, m\, \frac{\partial\alpha_R}{\partial m}$$

$$= 1 + \frac{\varkappa'}{\varkappa}\,\mu\,\frac{\partial \alpha_R}{\partial \mu}$$

$$= 1 + \frac{\varkappa'}{\varkappa}\,\alpha\,\beta_{tw}(\alpha)$$

With this we get a relation between β_{tw} and β_{cs} [31]

$$\beta_{cs}(\alpha) = \left[1 + \alpha\beta_{tw}(\alpha)\frac{\varkappa'(\alpha)}{\varkappa(\alpha)}\right]\beta_{tw}(\alpha) \tag{8.17}$$

The second term within the paranthesis leads to a term of the 3rd order in α, so that

$$\beta_{cs}(\alpha) = \beta_{tw}(\alpha) + O(\alpha^3) \quad , \quad \text{qed.}$$

(8.16) thus gives us

$$\beta_{tw}(\alpha) = \frac{2}{3}\left(\frac{\alpha}{\pi}\right) + \frac{1}{2}\left(\frac{\alpha}{\pi}\right)^2 + O(\alpha^3) \tag{8.18}$$

A comparison with (6.41) shows, moreover, that

$$\beta_f(\alpha) = \beta_{tw}(\alpha) + O(\alpha^2) = \beta_{cs}(\alpha) + O(\alpha^2)$$

We shall now use the results of our two-loop calculation to calculate β_{cs} to the order α^2. We found in (7.35) that

$$Z_3^{-1}(\alpha_0, m_0) = 1 + \frac{\alpha_0}{3\pi}\ln\frac{1}{i\gamma m_0^2 s_0} - \frac{\alpha_0^2}{4\pi^2}\ln\frac{1}{2i\gamma m_0^2 s_0} + O(\alpha_0^3)$$

This is now expressed by the physical mass m

$$m^2 = m_0^2 + \delta m^2$$

$$= m_0^2 + \frac{3\alpha_0 m^2}{2\pi}\left[\ln\frac{1}{i\gamma m^2 s_0} + \frac{5}{6}\right] + O(\alpha_0^2)$$

whereby, only the terms constant, linear and quadratic in α_0 need be taken into account:

$$Z_3^{-1}(\alpha_0, m) = 1 - \frac{\alpha_0}{3\pi}\ln[i\gamma s_0(m^2 - \delta m^2)] - \frac{\alpha_0^2}{4\pi^2}\ln\frac{1}{2i\gamma m^2 s_0} + O(\alpha_0^3)$$

$$= 1 - \frac{\alpha_0}{3\pi}\ln(i\gamma s_0 m^2) - \frac{\alpha_0^2}{4\pi^2}\left[\ln(i\gamma m^2 s_0) + C\right] + O(\alpha_0^3)$$

Therefore $\left(C := -\frac{5}{6} - \ln 2 \right)$

$$\beta_{cs}(\alpha_0) = \frac{m}{Z_3(\alpha_0, m)} \frac{\partial Z_3(\alpha_0, m)}{\partial m}$$

$$= -m\frac{\partial}{\partial m}\Big[-\frac{\alpha_0}{3\pi}\ln(i\gamma s_0 m^2) - \frac{\alpha_0^2}{4\pi^2}\{\ln(i\gamma m^2 s_0) + C\}$$
$$- \frac{1}{2}\left(\frac{\alpha_0}{3\pi}\right)^2 \ln^2(i\gamma s_0 m^2) + O(\alpha_0^3)\Big]$$

$$= \frac{2}{3}\frac{\alpha_0}{\pi} + \frac{1}{2}\left(\frac{\alpha_0}{\pi}\right)^2 + \frac{2}{3\pi}\cdot\frac{\alpha_0^2}{3\pi}\ln(i\gamma m^2 s_0) + O(\alpha_0^3) \tag{8.19}$$

Now we have to replace the bare coupling constant by the physical one with $\alpha = \alpha_0 Z_3(\alpha_0)$

$$\alpha = \alpha_0 Z_3(\alpha_0) = \alpha_0\left[Z_3^{-1}(\alpha_0)\right]^{-1}$$

$$= \alpha_0\left[1 - \frac{\alpha_0}{3\pi}\ln(i\gamma m^2 s_0) + O(\alpha_0^2)\right]^{-1}$$

$$= \alpha_0\left[1 + \frac{\alpha_0}{3\pi}\ln(i\gamma m^2 s_0)\right] + O(\alpha_0^3)$$

With

$$\alpha = \alpha_0 + \frac{\alpha_0^2}{3\pi}\ln(i\gamma m^2 s_0) + O(\alpha_0^3)$$

it follows from (8.19) for β_{cs} dependent on α

$$\beta_{cs}(\alpha) = \frac{2}{3}\frac{\alpha_0}{\pi} + \frac{1}{2}\left(\frac{\alpha_0}{\pi}\right)^2 + \frac{2}{3\pi}(\alpha - \alpha_0) + O(\alpha^3)$$

$$= \frac{2}{3}\left(\frac{\alpha}{\pi}\right) + \frac{1}{2}\left(\frac{\alpha}{\pi}\right)^2 + O(\alpha^3)$$

in complete accord with the result obtained from the study of the polarization function (8.16).

The above calculation is an example of the close relationship between polarization function and effective Lagrangian; in general $\ell = L/L^{(0)}$ in the limiting case of strong magnetic fields, contains the same information as $\Pi(q^2)$ does for large space-like arguments ($q^2 \gg m^2$). But ℓ can be perturbatively calculated from a small number of topologically different diagrams [4]: in the order $\alpha, \alpha^2, \alpha^3, \ldots$ one needs 1,2,10... graphs for Π; for ℓ, however, only 1,1,3,... .

In order to illustrate these facts, we show that $\Pi(q^2)$ coincides with ℓ for large space-like momenta with logarithmic exactness in the order α and α^2.
By means of dimensional regularization we get [32] for $q^2 \gg m_R^2$:

$$\Pi_R(q^2) \approx \frac{\alpha_R}{\pi}\left[-\frac{1}{3}\ln\frac{q^2}{\mu^2} + \frac{2m_R^2}{(-q^2)} + 2\left(\frac{m_R^2}{q^2}\right)^2 \ln\frac{q^2}{m_R^2} + \text{const}\right]$$

$$+\left(\frac{\alpha_R}{\pi}\right)^2\left[-\frac{1}{4}\ln\frac{q^2}{\mu^2} + \frac{3m_R^2}{q^2}\ln\frac{q^2}{\mu^2} - 6\left(\frac{m_R^2}{q^2}\right)^2 \ln\frac{q^2}{m_R^2}\ln\frac{q^2}{\mu^2} + \text{const}\right] + O(\alpha_R^3)$$

We substitute now $q^2 \to eB$ and multiply with $L^{(0)}$

$$\Pi_R(eB)\,\mathcal{L}^{(0)}(B)$$

$$\approx \frac{\alpha_R}{\pi}\left(-\frac{1}{2}B^2\right)\left[-\frac{1}{3}\ln\frac{eB}{\mu^2} - 2\frac{m_R^2}{eB} + 2\left(\frac{m_R^2}{eB}\right)^2 \ln\frac{eB}{m_R^2} + \text{const}\right] +$$

$$+\left(\frac{\alpha_R}{\pi}\right)^2\left(-\tfrac{1}{2}B^2\right)\left[-\tfrac{1}{4}\ln\frac{eB}{\mu^2}+3\frac{m_R^2}{eB}\ln\frac{eB}{\mu^2}-6\left(\frac{m_R^2}{eB}\right)^2\ln\frac{eB}{m_R^2}\cdot\right.$$

$$\left.\cdot\ln\frac{eB}{\mu^2}+\text{const}\right]+O(\alpha_R^3)$$

If we set $\mu = \kappa m$ and consider only the terms dominant for strong fields, then

$$\Pi(eB)\,\mathcal{L}^{(0)}(B)\approx\left(\frac{\alpha}{\pi}\right)\left(-\tfrac{1}{2}B^2\right)\left[-\tfrac{1}{3}\ln\frac{eB}{m^2}\right]$$

$$+\left(\frac{\alpha}{\pi}\right)^2\left(-\tfrac{1}{2}B^2\right)\left[-\tfrac{1}{4}\ln\frac{eB}{m^2}\right] \tag{8.20}$$

$$=\frac{\alpha B^2}{6\pi}\ln\frac{eB}{m^2}+\frac{\alpha^2B^2}{8\pi^2}\ln\frac{eB}{m^2}$$

On the other hand, it follows from (5.29) and (7.43) that

$$\lim_{\frac{eB}{m^2}\to\infty}\ell(B)\,\mathcal{L}^{(0)}(B)\equiv\lim_{\frac{eB}{m^2}\to\infty}\left(\mathcal{L}_R^{(1)}(B)+\mathcal{L}_R^{(2)}(B)\right)$$

$$=\frac{\alpha B^2}{6\pi}\ln\frac{eB}{m^2}+\frac{\alpha^2B^2}{8\pi^2}\ln\frac{eB}{m^2}+O(B^2)$$

Herewith, we have shown the equality of ℓ and Π in the given approximation.

Following Ritus [4], we shall now use the Callan-Symanzik equation (8.14)

$$\left[m^2\frac{\partial}{\partial m^2}+\bar{\beta}(\alpha)\left(\alpha\frac{\partial}{\partial\alpha}-1\right)\right]\ell_{R\infty}\left(\frac{eB}{m^2},\alpha\right)=0$$

with

$$\bar{\beta}(\alpha):=\tfrac{1}{2}\beta_{cs}(\alpha)$$

$$=\tfrac{1}{3}\left(\frac{\alpha}{\pi}\right)+\tfrac{1}{4}\left(\frac{\alpha}{\pi}\right)^2+O(\alpha^3) \tag{8.21}$$

to derive an improved form of L for strong fields. So we write

$$\bar{\beta}(\alpha) = \sum_{k=1}^{\infty} \bar{\beta}_k \left(\frac{\alpha}{\pi}\right)^k \quad ; \tag{8.22}$$

furthermore, it is known from the perturbation calculation in higher orders [42], that one can make the Ansatz for ℓ

$$\ell_{R,\infty}(z,\alpha) = 1 + \frac{\alpha}{\pi}(a_{10} + a_{11} z) + \sum_{n=2}^{\infty} \left(\frac{\alpha}{\pi}\right)^n \sum_{k=0}^{n-1} a_{nk} z^k \tag{8.23}$$

with

$$z := \ln \frac{eB}{\gamma \pi m^2}$$

Because of

$$m^2 \frac{\partial}{\partial m^2} = m^2 \frac{\partial z}{\partial m^2} \frac{\partial}{\partial z} = -\frac{\partial}{\partial z}$$

it follows that

$$\left[-\frac{\partial}{\partial z} + \bar{\beta}(\alpha)\left(\alpha \frac{\partial}{\partial \alpha} - 1\right)\right] \ell_{R\infty}(z,\alpha) = 0$$

Substitution of (8.22) and (8.23) gives:

$$\left[-\frac{\partial}{\partial z} + \sum_{l=1}^{\infty} \bar{\beta}_l \left(\frac{\alpha}{\pi}\right)^l \left(\alpha\frac{\partial}{\partial \alpha} - 1\right)\right]\left\{1 + \frac{\alpha}{\pi}(a_{10} + a_{11} z) + \sum_{n=2}^{\infty} \left(\frac{\alpha}{\pi}\right)^n \sum_{k=0}^{n-1} a_{nk} z^k\right\} = 0$$

After performing the derivatives, we get

$$-\frac{\alpha}{\pi} a_{11} - \sum_{n=2}^{\infty} \left(\frac{\alpha}{\pi}\right)^n \sum_{k=0}^{n-1} a_{nk} k z^{k-1}$$

$$+ \sum_{l=1}^{\infty} \bar{\beta}_l \left(\frac{\alpha}{\pi}\right)^l \left(\alpha\frac{\partial}{\partial \alpha} - 1\right) 1 + (a_{10} + a_{11} z) \sum_{l=1}^{\infty} \bar{\beta}_l \left(\frac{\alpha}{\pi}\right)^l \left(\alpha\frac{\partial}{\partial \alpha} - 1\right)\left(\frac{\alpha}{\pi}\right)$$

$$+ \sum_{l=1}^{\infty} \bar{\beta}_l \left(\frac{\alpha}{\pi}\right)^l \sum_{n=2}^{\infty} \left[\left(\alpha\frac{\partial}{\partial \alpha} - 1\right)\left(\frac{\alpha}{\pi}\right)^n\right] \sum_{k=0}^{n-1} a_{nk} z^k = 0$$

which can be simplified as

$$-\frac{\alpha}{\pi}a_{11} - \sum_{n=2}^{\infty}\left(\frac{\alpha}{\pi}\right)^n \sum_{k=0}^{n-1} a_{nk}\, k\, z^{k-1} - \underline{\sum_{l=1}^{\infty}\bar{\beta}_l\left(\frac{\alpha}{\pi}\right)^l}$$

$$+\sum_{l=1}^{\infty}\bar{\beta}_l\left(\frac{\alpha}{\pi}\right)^l \sum_{n=2}^{\infty}\left(\frac{\alpha}{\pi}\right)^n (n-1)\sum_{k=0}^{n-1} a_{nk}\, z^k \quad = 0 \tag{8.24}$$

A comparison of coefficients with α shows

$$a_{11} = -\bar{\beta}_1$$

while the vanishing of the terms constant with respect to z of the second and fourth term leads to

$$-\sum_{n=2}^{\infty}\left(\frac{\alpha}{\pi}\right)^n a_{n1} + \sum_{l=1}^{\infty}\bar{\beta}_l\left(\frac{\alpha}{\pi}\right)^l \sum_{n=2}^{\infty}\left(\frac{\alpha}{\pi}\right)^n (n-1)\, a_{n0} = 0$$

or to

$$\sum_{k=2}^{\infty}\left(\frac{\alpha}{\pi}\right)^k a_{k1} = \sum_{i=2}^{\infty}\sum_{l=1}^{\infty}\left(\frac{\alpha}{\pi}\right)^{l+i}(i-1)\, a_{i0}\,\bar{\beta}_l \tag{8.25}$$

If we now compare the coefficients of the various powers of α, we get

$$a_{k1} = \sum_{i=2}^{k-1}(i-1)\, a_{i0}\,\bar{\beta}_{k-i} = \sum_{i=1}^{k-1}(i-1)\, a_{i0}\,\bar{\beta}_{k-i}\,, \qquad k = 2, 3, 4, \ldots \tag{8.26}$$

Here the sum over i terminates after $i = k-1$, since it follows from (8.25) that $k = l+i$ and the smallest value for ℓ is exactly one. If we define $a_{00} := 1$, then we can generalize (8.26) to

$$a_{k1} = \sum_{i=0}^{k-1}(i-1)\, a_{i0}\,\bar{\beta}_{k-i}\,, \qquad k = 1, 2, 3, \ldots \tag{8.27}$$

this reproduces namely $a_{11} = -a_{00}\bar{\beta}_1 = -\bar{\beta}_1$ and simultaneously takes into account the underlined term (without the terms for $\ell = 1$) in (8.24). Here we have derived a first recursion formula for the coefficients in (8.23); a second can be obtained by a comparison of coefficients in (8.24) with respect to z^k, $k \geq 2$:

$$-\sum_{n=2}^{\infty}\left(\frac{\alpha}{\pi}\right)^n \sum_{k=0}^{n-1} a_{nk}\, k\, z^{k-1}$$

$$+\sum_{l=1}^{\infty}\bar{\beta}_l \left(\frac{\alpha}{\pi}\right)^l \sum_{n=2}^{\infty}\left(\frac{\alpha}{\pi}\right)^n (n-1)\sum_{k=0}^{n-1} a_{nk}\, z^k = 0$$

implies

$$\sum_{n=2}^{\infty}\left(\frac{\alpha}{\pi}\right)^n a_{nk}\, k = \sum_{l=1}^{\infty}\sum_{i=2}^{\infty}\left(\frac{\alpha}{\pi}\right)^{l+i}\bar{\beta}_l\,(i-1)\, a_{i,k-1}$$

so that a comparison of coefficients with respect to the orders of magnituds of α

$$k a_{nk} = \sum_{i=k}^{n-1}(i-1)\, a_{i,k-1}\,\bar{\beta}_{n-i}$$

results.

As a consequence of the renormalization group equation, the expansion coefficients a_{nk} must satisfy the following recursion relations

$$\begin{aligned} a_{00} &= 1 \\ a_{n1} &= \sum_{i=0}^{n-1}(i-1)\, a_{i0}\,\bar{\beta}_{n-i} \\ k a_{nk} &= \sum_{i=k}^{n-1}(i-1)\, a_{i,k-1}\,\bar{\beta}_{n-i}\,, \quad k \geq 2\,. \end{aligned} \tag{8.28}$$

With the aid of this equation, we can show that in (8.23), to all orders in α, the coefficient of the highest power of $\ln(eB/m^2)$ (this is $a_{11} = -\beta_1$ and $a_{n,n-1}$, $n = 2,3,\ldots$) is determined by $\bar{\beta}_1$ and $\bar{\beta}_2$ alone. For this we use

$$k a_{Nk} = \sum_{i=k}^{N-1}(i-1)\, a_{i,k-1}\,\bar{\beta}_{N-i}$$

and calculate

$$
\begin{aligned}
(n-1)\,a_{n,n-1} &\underset{N=n}{=} \sum_{i=n-1}^{n-1} (n-2)\,a_{n-1,n-2}\,\bar{\beta}_{n-n+1} \\
&= \bar{\beta}_1\,(n-2)\,a_{n-1,n-2} \\
&\underset{N=n-1}{=} \bar{\beta}_1 \sum_{i=n-2}^{n-2} (n-3)\,a_{n-2,n-3}\,\bar{\beta}_{n-1-n+2} \\
&= \bar{\beta}_1^{2}\,(n-3)\,a_{n-2,n-3} \\
&\underset{N=n-2}{=} \bar{\beta}_1^{2}\,(n-4)\,a_{n-3,n-4}\,\bar{\beta}_{n-2-n+3} \\
&= \bar{\beta}_1^{3}\,(n-4)\,a_{n-3,n-4} \\
&\underset{N=n-3}{=} \bar{\beta}_1^{3}\,(n-5)\,a_{n-4,n-5}\,\bar{\beta}_{n-3-n+4} \\
&= \bar{\beta}_1^{4}\,(n-5)\,a_{n-4,n-5} \\
&\;\vdots \\
&= \bar{\beta}_1^{(n-2)}\big(n-(n-1)\big)\,a_{n-(n-2),\,n-(n-1)} \\
&= \bar{\beta}_1^{(n-2)}\,a_{21} \\
&= \bar{\beta}_1^{(n-2)}\left(-a_{00}\,\bar{\beta}_2\right) \\
&= -\,\bar{\beta}_2\,\bar{\beta}_1^{(n-2)}
\end{aligned}
$$

So

$$
a_{n,n-1} = -\,\frac{\bar{\beta}_2\,\bar{\beta}_1^{(n-2)}}{n-1} \qquad (8.29)
$$

which, as claimed, depends only on $\bar{\beta}_1$ and $\bar{\beta}_2$.

Analogously, one could calculate $a_{n,n-2}$ and would have then in each order of α the coefficients to the second-highest power of $\ln(eB/m^2)$, etc.

We now define

$$\left(\frac{\alpha}{\pi}\right)^r L_r := \sum_{n=r+1}^{\infty} \left(\frac{\alpha}{\pi}\right)^n a_{n,n-r}\, z^{n-r}\,, \quad r = 1,2,\ldots \tag{8.30}$$

The simplest examples are

$$\left(\frac{\alpha}{\pi}\right) L_1 = \left(\frac{\alpha}{\pi}\right)^2 a_{21} z + \left(\frac{\alpha}{\pi}\right)^3 a_{32} z^2 + \left(\frac{\alpha}{\pi}\right)^4 a_{43} z^3 + \left(\frac{\alpha}{\pi}\right)^5 a_{54} z^4 + \ldots$$

$$\left(\frac{\alpha}{\pi}\right)^2 L_2 = \left(\frac{\alpha}{\pi}\right)^3 a_{31} z + \left(\frac{\alpha}{\pi}\right)^4 a_{42} z^2 + \left(\frac{\alpha}{\pi}\right)^5 a_{53} z^3 + \ldots$$

$$\left(\frac{\alpha}{\pi}\right)^3 L_3 = \left(\frac{\alpha}{\pi}\right)^4 a_{41} z + \left(\frac{\alpha}{\pi}\right)^5 a_{52} z^2 + \ldots$$

$$\left(\frac{\alpha}{\pi}\right)^4 L_4 = \left(\frac{\alpha}{\pi}\right)^5 a_{51} z + \ldots$$

So L_1 is the sum of the highest powers of z in every order of α, L_2, the sum of the second-highest orders, etc. With (8.30), we can also write the series for $\ell_{R\infty}$ as

$$\ell_{R\infty} = 1 - x + \sum_{n=1}^{\infty} \left(\frac{\alpha}{\pi}\right)^n \left(L_n(x) + a_{n0}\right) \tag{8.31}$$

with

$$x := \frac{\alpha}{\pi}\bar{\beta}_1 z = \frac{\alpha}{\pi}\bar{\beta}_1 \ln\frac{eB}{\gamma\pi m^2}$$

The expansion in α no longer signifies an expansion according to the number of photons; rather $\left(\frac{\alpha}{\pi}\right)^n L_n$ is the sum of those logarithms whose power is smaller by n than the largest possible in the corresponding order of α.

With (8.29), we can sum up the leading logarithms to any order in closed form

$$
\begin{aligned}
L_1 &= \left(\frac{\alpha}{\pi}\right)^{-1} \sum_{n=2}^{\infty} \left(\frac{\alpha}{\pi}\right)^{n} a_{n,n-1}\, z^{n-1} \\
&= \sum_{n=2}^{\infty} \left(\frac{\alpha}{\pi}\right)^{n-1} (n-1)^{-1}\, \bar{\beta}_1^{\,n-2}\, z^{n-1} \left(-\bar{\beta}_2\right) \\
&= \sum_{n=2}^{\infty} \left(\frac{\alpha}{\pi}\right)^{n-1} (n-1)^{-1}\, z^{n-1} \left(-\frac{\bar{\beta}_2}{\bar{\beta}_1}\right) \\
&= -\frac{\bar{\beta}_2}{\bar{\beta}_1} \sum_{k=1}^{\infty} \frac{1}{k} x^k \\
&= \frac{\bar{\beta}_2}{\bar{\beta}_1} \ln(1-x)\,, \qquad x \in (-1,+1)\,.
\end{aligned}
\tag{8.32}
$$

Here, we have used the series

$$
\ln(1-x) = -\sum_{k=1}^{\infty} \frac{1}{k} x^k\,, \quad x \in (-1,+1)\,.
$$

One can perform an analogous calculation for L_2 too, but the result would contain the coefficients a_{20} and $\bar{\beta}_3$ not calculated here; the same applies for the higher L_n. If we consider only L_1, then (8.31) can be simplified to

$$
\ell_{R\infty} = 1 - x + \frac{\alpha}{\pi} L_1(x) + \frac{\alpha}{\pi} a_{10} = 1 - x + \frac{\alpha}{\pi}\frac{\bar{\beta}_2}{\bar{\beta}_1} \ln(1-x) + \frac{\alpha}{\pi} a_{10}
$$

With the help of this representation, we can now evaluate the domain of validity of our improved, asymptotic form.

If we leave out L_1, then, with logarithmic accuracy,

$$
\ell_{R\infty} = 1 - x \quad .
$$

This formula is only applicable if the deviation from the 'unperturbed' Lagrangian $L^{(o)}$ (i.e. form $\ell_{R\infty}=1$) is small, i.e. for

$$x = \frac{\alpha}{\pi}\,\bar{\beta}_1 \ln\frac{eB}{\gamma\pi m^2} \ll 1 .$$

If L_1 is also taken into consideration, then its contribution must be small with respect to (1-x), i.e.

$$\frac{\alpha}{\pi}\,|\ln(1-x)| \ll |1-x|$$

must be valid.

Because of $x < 1$, this means

$$-\frac{\alpha}{\pi}\ln(1-x) = \frac{\alpha}{\pi}\ln\frac{1}{1-x} \ll (1-x)$$

or

$$\left(\frac{\alpha}{\pi}\right)\frac{\ln\frac{1}{1-x}}{(1-x)} \ll 1 . \qquad (8.33)$$

This means that the improved asymptotic form for L is valid in an enlarged region of x; $x \ll 1$ must not necessarily apply and larger x's are possible, buth they must still satisfy (8.33).

So, neglecting L_2 and higher terms, we get

$$\frac{\mathcal{L}_{R\infty}}{\mathcal{L}_R^{(0)}} \equiv \ell_{R\infty} = 1-x+\frac{\alpha}{\pi}\frac{\bar{\beta}_2}{\bar{\beta}_1}\ln(1-x)+\frac{\alpha}{\pi}a_{10}$$

$$= 1-\left(\frac{\alpha}{\pi}\right)\bar{\beta}_1\ln\frac{eB}{\gamma\pi m^2}$$

$$+\left(\frac{\alpha}{\pi}\right)\frac{\bar{\beta}_2}{\bar{\beta}_1}\ln\left[1-\frac{\alpha}{\pi}\bar{\beta}_1\ln\frac{eB}{\gamma\pi m^2}\right]+\left(\frac{\alpha}{\pi}\right)a_{10}$$

with $\bar{\beta}_1 = \frac{1}{3}$ and $\bar{\beta}_2 = \frac{1}{4}$ according to (8.21). The coefficient a_{10} results from (5.29)

$$\mathcal{L}_R^{(1)}\left(\frac{eB}{m^2} \to \infty\right) = \frac{\alpha B^2}{6\pi}\left[\ln \frac{eB}{m^2} + 12\zeta'(-1) - 1 + \ln 2\right]$$

$$= \frac{\alpha B^2}{6\pi}\left[\ln \frac{eB}{\gamma\pi m^2} + \frac{6}{\pi^2}\zeta'(2)\right]$$

$$\equiv \ell_{R\infty}\cdot\left(-\frac{1}{2}B^2\right)$$

and (8.23) to

$$a_{10} = -\frac{2}{\pi^2}\zeta'(2)$$

It thus follows, for the improved asymptotic form

$$\mathcal{L}_{R\infty}(B) = -\frac{1}{2}B^2 + \frac{\alpha B^2}{6\pi}\left[\ln\frac{eB}{\gamma\pi m^2} + \frac{6}{\pi^2}\zeta'(2)\right]$$

$$-\frac{\alpha B^2}{2\pi}\,\frac{3}{4}\ln\left[1 - \frac{\alpha}{3\pi}\ln\frac{eB}{\gamma\pi m^2}\right]$$

or

$$\mathcal{L}_{R\infty}(B) = -\frac{1}{2}B^2 + \frac{\alpha B^2}{6\pi}\left[\ln\frac{eB}{m^2} + 12\zeta'(-1) - 1 + \ln 2\right]$$

$$-\frac{\alpha B^2}{2\pi}\,\frac{3}{4}\ln\left[1 - \frac{\alpha}{3\pi}\ln\frac{eB}{\gamma\pi m^2}\right] \qquad (8.34)$$

The third term apparently contains a singularity at $B = B_D$ with

$$B_D := \gamma\pi\frac{m^2}{e}\exp\left(\frac{3\pi}{\alpha}\right) \ggg \frac{m^2}{e} = 4.4 \cdot 10^{13} \text{ Gauss},$$

in the proximity of which (8.34) apparently is not valid. Field strengths of the size of B_D do not, however, have any physical meaning, since they are still extremely large, even compared to the 'critical field strength' m^2/e.

To conclude, we want to return for a moment to the case of the pure electric field. For $eE \gg m^2$ we get from (5.29), or for the second term in (8.34), by means of the substitution $B \to \frac{1}{i} E$

$$\mathrm{Re}\, \mathcal{L}_R^{(1)}\left(\frac{eE}{m^2} \to \infty\right) = -\frac{\alpha E^2}{6\pi}\left[\ln\frac{eE}{m^2} + 12\zeta'(-1) - 1 + \ln 2\right] \tag{8.35a}$$

$$\mathrm{Im}\, \mathcal{L}_R^{(1)}\left(\frac{eE}{m^2} \to \infty\right) = \frac{\alpha}{12} E^2 \tag{8.35b}$$

We now establish that the radiative corrections to an arbitrary high order in α contained in L_1 do not correct the imaginary part of the one-loop effective Lagrangian; the last term in (8.34) leads, in fact, to

$$\ln\left[1 - \frac{\alpha}{3\pi}\ln\frac{(-ieE)}{\gamma\pi m^2}\right] = \ln\left[1 - \frac{\alpha}{3\pi}\left\{\ln\frac{eE}{\gamma\pi m^2} - i\frac{\pi}{2}\right\}\right]$$

$$\approx \ln\left[1 - \frac{\alpha}{3\pi}\ln\frac{eE}{\gamma\pi m^2}\right]$$

$$\text{for} \quad eE \gg m^2 .$$

An imaginary part could only appear in this expression for

$$\frac{\alpha}{3\pi}\ln\frac{eE}{\gamma\pi m^2} > 1$$

which, on the one hand, would contradict the prerequisite $x < 1$ in (8.32) and on the other, would require unrealistically large field strengths $E \ggg m^2/e$. So (8.53b) undergoes no radiative corrections in the framework of the approximation of (8.34).

This example clearly shows the usefulness of the renormalization group equations. Knowing only the coefficients β_1 and β_2 (note that the latter could also be obtained by calculating $\Pi(q^2)$ at the two-loop level, where the case of B = 0 is completely sufficient, of course), we were able to sum up the leading logarithms of every order of perturbation theory and to thus improve our former one-loop calculation by extending its range of validity with respect to the strength of the field. This leads to the typical ln (ln B) term in (8.34).

(9) Applications and Discussion

In this final section we want to make contact with the work of other authors and, as a further example illustrating the non-trivial vacuum of QED, discuss the corrections of the Coulomb law due the presence of the quantized fermions.

Up to now, we have always restricted our calculations to the case of constant electric or magnetic fields. However, it can be shown that the leading terms for strong fields, i.e., those of order $B^2 \ln B$ or $E^2 \ln E$, are the same even if the fields are not constant. In a heuristic way this can be shown as follows (for a more rigorous discussion, see [57] and [58]): One starts from the Maxwell Lagrangian

$$\mathcal{L} = -\frac{1}{4} F_{\mu\nu} F^{\mu\nu} \tag{9.1}$$

and scales the electromagnetic coupling e out of the fields:

$$A_\mu \rightarrow \frac{1}{e} A_\mu \tag{9.2}$$

giving now

$$\mathcal{L} = -\frac{1}{4e^2} F_{\mu\nu} F^{\mu\nu} \tag{9.3}$$

Note that in the complete interacting QED Lagrangian this is the only term containing e, because the vertex now simply reads $\bar{\psi}A\!\!\!/\psi$ instead of e $\bar{\psi}A\!\!\!/\psi$. The next step is to "renormalization-group-improve" (9.3) by replacing e with the running coupling constant $e(\mu)$ to first order in α. This function is determined as the solution of (8.7) when including only the $O(\alpha)$-term in the β-function. For $e^2(\mu) \equiv 4\pi\,\alpha(\mu)$, one gets [12] the scaling equation

$$e^2(\mu) = \frac{e^2(\mu_0)}{1 - \frac{e^2(\mu_0)}{6\pi^2} \ln \frac{\mu}{\mu_0}} \tag{9.4}$$

describing how the coupling changes when the scale μ varies; hereby $e^2(\mu_0)$ is an integration constant. In our applications, where the fields are sufficiently strong so that fermionic masses are negligible, the length or mass scale is set by the magnitude of $F^2 \equiv F_{\mu\nu} F^{\mu\nu}$. Therefore we replace μ^4 in (9.4) by F^2 (recall dim F = $(\text{mass})^2$!) to obtain

$$e^2(F^2) = \frac{e^2(\mu_0)}{1 - \frac{e^2(\mu_0)}{24\pi^2} \ln \frac{F^2}{\mu_0^4}} \tag{9.5}$$

with an arbitrary reference mass μ_0. The last step is to replace e^2 in (9.3) by the field dependent running coupling constant $e^2(F^2)$:

$$\mathcal{L}_{eff} = -\frac{1}{4e^2(F^2)} F_{\mu\nu} F^{\mu\nu} \tag{9.6}$$

$$= -\frac{1}{4e^2(\mu_0)} F_{\mu\nu}F^{\mu\nu}\left[1-\frac{e^2(\mu_0)}{24\pi^2}\ln\frac{F^2}{\mu_0^4}\right]$$

The one-loop part is (we scale back $e^2(\mu_0) \equiv e^2$ into the fields)

$$\mathcal{L}^{(1)} = +\frac{1}{4}F_{\mu\nu}F^{\mu\nu}\cdot\frac{e^2}{24\pi^2}\ln\frac{e^2F^2}{\mu_0^4}$$

$$= \frac{1}{2}(\vec{B}^2-\vec{E}^2)\frac{e^2}{24\pi^2}\ln\frac{\vec{E}^2-\vec{B}^2}{\mu_0^4} + O(F^2) \qquad (9.7)$$

yielding for a pure magnetic field, for instance,

$$\mathcal{L}^{(1)}(B) = \frac{(eB)^2}{24\pi^2}\ln\frac{eB}{\mu_0^2} \qquad (9.8)$$

After making the special choice $\mu_0 = m$ this is up to terms of order $B^2 \ln(B)$ the same as our old equation (5.29) which was derived in section (5) for constant fieldsonly. At no point of the above "derivation" of (9.7) did we have to demand the fields to be constant; thus we may assume that (9.7) is correct to order $F^2 \ln F$ for arbitraryly varying fields. This is important information, because, as already mentioned, exact formulae for $L^{(1)}$ are known only for a very limited class of fields.

The above manipulations can be justified by noting that the ansatz (9.6) leads to the correct trace anomaly of the energy momentum tensor; for a thorough discussion of this point, see Pagels and Tomboulis [57].

Now that we have established the Lagrangian (9.7) for strong but otherwise arbitrary fields, we can set up the generalized

Maxwell equations and try to solve them for a given source distribution. In general, they are of the form

$$\frac{\delta}{\delta A_\mu(x)} \int d^4x' \left\{ \mathcal{L}^{(0)} + \mathcal{L}^{(1)} - J_\mu A^\mu \right\} = 0 \qquad (9.9)$$

with $L^{(1)}$ given by (9.7); hereby J_μ is a classical source charge current. Of particular interest are the problems of electrostatics where we have $J_\mu(x) = J_0(\vec{x})\, \delta_{\mu 0}$, $\vec{B} = 0$ and $\vec{E}(x) = \vec{E}(\vec{x}) = -\vec{\nabla} A^0(\vec{x})$. This leads us to evaluate

$$\frac{\delta}{\delta A^0(\vec{x})} \int d^3x' \left\{ \frac{1}{2} |\vec{\nabla} A^0|^2 \left[1 - \frac{e^2}{12\pi^2} \ln \frac{e|\vec{\nabla} A^0|}{m^2} \right] - J_0 A^0 \right\} = 0 \qquad (9.10)$$

Note that changing μ_0 to μ_0' in (9.7) gives rise only to a subdominant $O(B^2)$-term; therefore we may set $\mu_0 = m$ from now on.

The physical content of the variational problem (9.10) is easily visualized in terms of the dielectric "constant" of the vacuum

$$\varepsilon(\vec{E}) = 1 - \frac{e^2}{12\pi^2} \ln \frac{e|\vec{E}|}{m^2} \qquad (9.11)$$

and the displacement vector

$$\vec{D} \equiv \varepsilon(\vec{E})\, \vec{E} \,. \qquad (9.12)$$

In terms of these quantities, (9.10) simply reads

$$\operatorname{div} \vec{D} = J_0 \qquad (9.13)$$

This is a well-known equation from electrostatics of polarizable media. Looking back at the microscopic origin of $\varepsilon(\vec{E})$, we see that the effect of the vacuum fluctuations of the electron field is such that the vacuum responds to an external electric field as if it were some sort of crystal which possesses a field dependent dielectric "constant". Obviously, Maxwell's equations become non-linear due to the logarithm in (9.11).

To summarize, we can say that in deriving (9.13) with (9.11) and (9.12), we have solved the problem of finding the non-linear generalizations of Maxwell's equations - the non-linearities being caused by the electrons, which are hidden from direct observation but which influcncc the dynamics of the A_μ-field - for strong and static, but otherwise arbitrary, electrical field.
To get some insight into the effects produced by the second term in eq. (9.11), let us look at a specific example. We consider the case where J_0 contains only a single isolated charge Q at $\vec{x} = 0$ (together with a compensating spherical shell of charge -Q at infinity) [56,57]:

$$J_0(\vec{x}) = Q\,\delta(\vec{x}) \tag{9.14}$$

Making the spherically symmetric ansatz

$$\vec{D} = \frac{Q}{4\pi r^2}\hat{r}\ ,\ \vec{E} = \frac{Q(r)}{4\pi r^2}\hat{r}\ ,\ r \equiv |\vec{r}|\ ,\ \hat{r} \equiv \vec{r}/r \tag{9.15}$$

the equation (9.13) is solved provided that the function Q(r) is a solution of the transcendental equation

$$Q = Q(r)\ \mathcal{E}\left(\frac{e\,Q(r)}{4\pi r^2}\right) \tag{9.16}$$

The physical interpretation of Q(r) is that it is the charge lying within a sphere of radius r centered at $\vec{x} = 0$. The value Q(r) is always <u>larger</u> than Q because the vacuum polarization effects screen the charge. If we let $r \to \infty$, Q(r) approaches the (macroscopically) observed charge Q [37]. We thus got an implicit equation for the modification of Coulomb's law by the electron fluctuations:

$$E(r) = \frac{Q(r)}{4\pi r^2} \tag{9.17}$$

We stress that this equation is derived for strong fields, and hence short distances r, only. ("Strong" and "short" refer to the scale set by m^2 and m^{-1}, respectively). Of course, for extremely high field strengths, the one-loop approximation becomes invalid because the inequality $x \ll 1$ (cf. section (8)) does not hold any longer. In this region it becomes advantageous to use the renormalization group improved Lagrangian (8.34) because of its greater domain of validity.

We have now investigated the modification of Coulomb's law at very short distances; the contrary limiting case of larger r, however, can also be treated in a relatively simple way. Equivalently, we will ask for the effective Lagrangian for <u>weak</u>, but otherwise arbitrary, fields. In the weak field limit, $e^2 F_{\mu\nu} F^{\mu\nu}/m^4$ becomes small due to the smallness of α. This implies that the contributions of

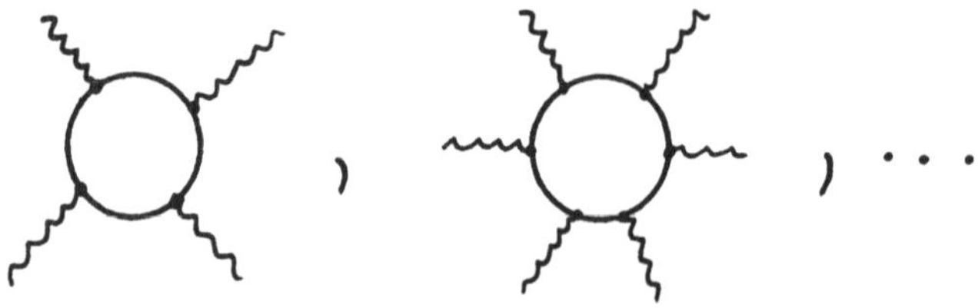

and those of higher order diagrams contained in $W^{(1)}$ (see appendix G) can be neglected relative to the diagram with only two vertices:

(9.18)

(The wavy lines are interactions with the external field, no photons!) Following appendix G, this leads us to the expression (wf = weak field)

$$W^{(1)}_{wf}[A] \equiv \int d^4x\, \mathcal{L}^{(1)}_{wf} = \frac{1}{2}\int d^4x\, d^4y\; A^{\mu}(x)\, \Pi_{\mu\nu}(x,y)\, A^{\nu}(y) \tag{9.19}$$

for the weak field limit of the one-loop effective action where $\Pi_{\mu\nu}$ is nothing but the order-e^2 polarization tensor (without external field) derived in section (4). In momentum space it is given by (4.34):

$$\Pi_{\mu\nu}(k) = (g_{\mu\nu}k^2 - k_\mu k_\nu)\,\Pi(k^2) \tag{9.20a}$$

$$\Pi(k^2) = -\frac{\alpha}{3\pi}k^2 \int_{4m^2}^{\infty} \frac{dt}{t}\, \varrho(t)\, \frac{1}{k^2 + t - i\varepsilon} \tag{9.20b}$$

$$\varrho(t) = \left(1 + \frac{2m^2}{t}\right)\left(1 - \frac{4m^2}{t}\right)^{1/2} \tag{9.20c}$$

As a consequence of the particular tensor structure of (9.20a), $W^{(1)}_{wf}$ can be written in terms of $F_{\mu\nu}$ only and therefore is gauge invariant, as it must be. Substituting the Fourier transform of (9.20) into (9.19) and adding the classical Maxwellian term, we get for the weak field effective Lagrangian

$$\mathcal{L}^{eff}_{wf} = -\frac{1}{4} F_{\mu\nu}(x)\left[1+\frac{\alpha}{3\pi}\,\Box\int_{4m^2}^{\infty}\frac{dt}{t}\,\frac{\rho(t)}{t-\Box}\right]F^{\mu\nu}(x) \tag{9.21}$$

where, as usual, $\Box \equiv -\partial_t^2 + \vec{\nabla}^2$. This is the weak field counterpart of eq. (9.6). There are two major differences between these two Lagrangians:

(i) The equations of motion derived from (9.6), i.e., eq. (9.13) in the static case, are non-linear due to the logarithm appearing in $\varepsilon(E)$. The equations of motion resulting from (9.21) for the weak field case are linear, because L^{eff}_{wf} is quadratic in the fields.

(ii) The strong field Lagrangian (9.6) is a local function of $F_{\mu\nu}(x)$; the weak field Lagrangian is non-local due to the $\Box$ -operator in the square bracket in (9.21).

Next, let us apply (9.21) to the Coulomb problem; specializing to the static case yields the field equation

$$\frac{\delta}{\delta A^0(\vec{x})}\int d^3x' \left\{\frac{1}{2}\vec{\nabla}A^0\cdot\left[1+\frac{\alpha}{3\pi}\vec{\nabla}^2\int_{4m^2}^{\infty}\frac{dt}{t}\,\frac{\rho(t)}{t-\vec{\nabla}^2}\right]\vec{\nabla}A^0 - A^0 J_0\right\} = 0 \tag{9.22}$$

For $J_0(\vec{x})$ we assume two point charges with separation r:

$$J_0(\vec{x}) = Q\{\delta(\vec{x}-\vec{x}_1)-\delta(\vec{x}-\vec{x}_2)\}\ ,\quad |\vec{x}_1-\vec{x}_2| \equiv r \tag{9.23}$$

The variation in (9.22) then gives the equation of motion

$$\mathcal{D}\,\vec{\nabla}^2 A^0(\vec{x}) = -J_0(\vec{x}) \tag{9.24}$$

where

$$\mathcal{D} \equiv 1 + \frac{\alpha}{3\pi}\,\vec{\nabla}^2 \int_{4m^2}^{\infty} \frac{dt}{t}\,\frac{\rho(t)}{t-\vec{\nabla}^2} \tag{9.25}$$

Making use of (9.24) and the position space representation of the resolvent $(t-\vec{\nabla}^2)^{-1}$, one easily calculates the potential energy $V = -\int d^3x\, L_{wf}^{eff}(A^0)$ associated with the interaction of the two point charges. One finds

$$V(r) = -\frac{Q^2}{4\pi}\left[\frac{1}{r} + \frac{\alpha}{3\pi}\int_{4m^2}^{\infty}\frac{dt}{t}\,\rho(t)\,\frac{e^{-\sqrt{t}\,r}}{r}\right] \tag{9.26}$$

The second term in the above bracket is the well-known Uehling correction of the Coulomb potential. Eq. (9.26) was derived in the weak field limit and thus should be valid at large distances. Obviously, the quantum corrections vanish for $r \to \infty$ and the classical $1/r$-behaviour is recovered. Because the equations of motion (9.24) are linear, (9.26) takes the form of a superposition of Yukawa potentials.

To summarize the two limiting cases discussed above [59], we can say that the QED vacuum behaves as a linear but spatially non-local medium at large distances (i.e., for weak fields), whereas it behaves as a local but non-linear medium at short distances (i.e., for strong fields). The calculation of an effective Lagrangian for arbitrary, but weak, fields is a relatively simple matter, because only the diagram (9.18) has to be considered. Larger calculational effort is needed for strong fields; in this case, all orders in the coupling of the fermion loop of the external field must be taken into account. Even in the one- or two-loop approximation this summation over an infinite set of diagrams can be done in closed form only for very special types of fields, constant fields, as considered here, or laser fields (i.e., plane waves) [3], for instance. Our computations in sections (5) to (8) are to be understood as a modest attempt to understand the QED vacuum at high field strengths or, equivalently, at short distances. As we have seen, the simplest non-linear, local quantum correction to the Maxwell Lagrangian arises from the diagram (5.4) which correctly describes the physics for strong fields as long as $L^{(1)}(\vec{E},\vec{B})$ is still well below $L^{(o)}(\vec{E},\vec{B})$, i.e., as long as the quantum corrections are small relative to the classical contribution. The point where they would become equal is the well-known Landau singularity [12]. It appears when the denominator of eq. (9.5) for the running coupling (to one-loop order) vanishes. Fields of this intensity, however, are of no physical significance, because they are many orders of magnitude stronger than the "critical field strength" m^2/e, which is already well beyond the magnitude of all laboratory physics. (See the discussion following eq. (8.34)). If one wants to proceed a little further towards the Landau singularity, one may use the renormalization group improved Lagrangian derived in section (8). Equation (8.34) is still valid for values of B too large for the simple $L^{(1)}(B)$ to be applicable [4]. Also this improved equation has the typical non-linear but local form, of course. (We assume, inspite of their derivation for constant fields, these equations to also be valid for slowly varying fiels).

Another type of correction to the Heisenberg-Euler Lagrangian which must be considered is the modification due to the quantized radiation field, the lowest order one, $L^{(2)}$, being represented by the two-loop diagram (7.6). As was shown in section (7), for constant $\vec{E}$- or $\vec{B}$-fields its contribution is about two orders of magnitude smaller than that of $L^{(1)}$; this should remain qualitatively true also for, at least slowly (on the scale of m^{-1}), varying fields. Again $L^{(2)}$ is a non-linear, but local, function of the fields. In the weak field domain these corrections are taken into account (for otherwise arbitrary fields) by using a radiatively corrected polarization tensor in (9.19). The result is again a non-local Lagrangian quadratic in the fields, i.e., the equations of motion remain linear.

Having now discussed in some detail the use of effective Lagrangians to characterize the vacuum of QED, let us finally briefly consider related problems in quantum chromodynamcis (QCD), which is believed to be the correct theory of strong interactions [60,12,51,53]. One of the salient features of this theory is asymptotic freedom, i.e., the fact that the coupling constant decreases when the mass scale is increased or the length scale is decreased. This behaviour is caused by the fact that the sign of the leading term in the QCD β-function is changed compared to the QED β-function. To lowest order, the coupling g scales as

$$g^2(\mu) = \frac{g^2(\mu_0)}{1 + b_0 g^2(\mu_0) \ln \frac{\mu}{\mu_0}} \tag{9.27}$$

with (N_f = number of quark flavours)

$$b_0 = \frac{1}{8\pi}\left(11 - \frac{2}{3} N_f\right). \tag{9.28}$$

This is the QCD analogue to eq. (9.4). However, for N_f being sufficiently small, b_0 is a positive constant so there is a crucial difference between (9.27) and (9.4): the sign of the quantum corrections in the denominators is such that the coupling <u>increases</u> in QED for increasing mass scale μ, but <u>decreases</u> in QCD for increasing μ. Or, stated in terms of a characteristic length scale μ^{-1}: in QED the coupling decreases when the distance becomes larger, but it increases in QCD.

Just as in QED, we can derive effective Lagrangians for this theory; they are the formal expression for the interactions of an external color field with the quark/gluon vacuum fluctuations. A new feature appearing here is the possibility of gluon-gluon interactions; this means that in one loop calculations not only fermion, but also gluon loops (and ghost loops) must be considered [61]. To get a first understanding of the effects associated with these Lagrangians, we can use the classical Yang-Mills Lagrangian [12 ,25]

$$\mathcal{L}^{(0)} = -\frac{1}{4g^2} F^a_{\mu\nu} F^{a\mu\nu} \tag{9.29}$$

with the coupling scaled out and replace g^2 by the function $g^2(F^2)$ obtained by substituting F^2 for μ^4 in (9.27). (cf. the corresponding manipulations in QED). The result reads [62]

$$\mathcal{L}_{eff} = \frac{1}{2g^2}(E^2 - B^2)\left[1 + \frac{1}{4} b_0 g^2 \ln \frac{E^2 - B^2}{\mu_0^4}\right]$$
$$= \frac{1}{8} b_0 (E^2 - B^2) \ln \frac{E^2 - B^2}{e\varkappa^2} \tag{9.30}$$

with

$$\varkappa^2 \equiv \frac{\mu_0^4}{e} \exp\left[-\frac{4}{b_0 g^2(\mu_0)}\right] \tag{9.31}$$

and $E^2 \equiv \vec{E}^a \cdot \vec{E}^a$ and similar for B^2. It is easy to show that the quantity κ defined in (9.31) is renormalization group invariant, i.e., it does not change when varying μ_0. Hence κ is a physical, i.e., observable, quantity. As a function of $E^2 - B^2$, L_{eff} has qualitatively the following behaviour

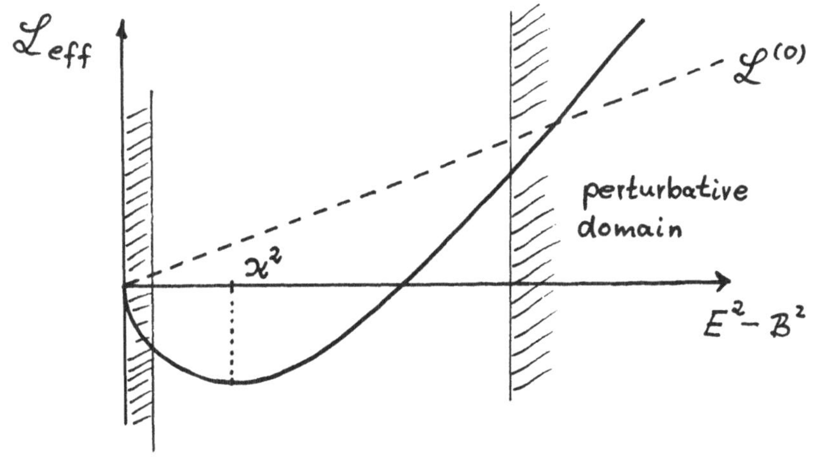

Now let us ask where we can trust the above curve. We know that in QCD, perturbation theory is valid at short distances, i.e., for strong fields (on the scale of κ). In this region the deviation of L_{eff} from the classical Young-Mills Lagrangian is small. Following the arguments of Adler [62,63], there is another region where (9.30) should be correct; this is the domain near the origin, because for $E^2 - B^2$ very small, i.e, μ in (9.27) very small, $g^2(\mu)$ is again small (but negative!!) and perturbation theory (in this case a one loop calculation) should be possible. If we accept this argument, we know that L_{eff} has negative slope near the origin but positive slope for strong fields. Despite the fact that (9.31) becomes untrustworthy in the intermediate region, we can conclude that L_{eff} must possess a minimum for a non-vanishing value of $E^2 - B^2$ (its continuity assumed). One may therefore assume that (9.30) interpolates qualitatively correctly in the strong coupling regime $E^2 - B^2 \approx \kappa^2$. It is therefore sensible to use (9.30) as a model of the QCD vacuum for arbitrary fields. Having an effective Lagrangian, it is natural to set up the modified equations of motion for the color fields (the analogue of (1.6)) and to try to solve them in presence of external sources, just as we did in QED when investigating the modification of Coulomb's law. What one would like to explore using this method is the question of color confinement; referring to the quarks, this means that the static potential between two quarks should rise at least linearly with the distance because then the energy to separate two quarks would be infinite. Surely, the discussion of confinement purely in terms of the potential of static color sources (i.e., infinitely heavy quarks) is not the whole story, but it is a first step in understanding the non-perturbative effects in QCD.

The general program of the so-called leading-log model [62,63] based upon (9.30) is quite similar to what we did in QED starting from (9.7). One sets up the static equations of motion for the fields coupled to two static point color charges corresponding to a quark/anti-quark pair, solves them and calculates the interaction energy of the quarks as a function of their distance. However, due to the non-linear character of the equations, this is a highly non-trivial task which requires sophisticated numerical methods. The details of this procedure are too complicated to be reported here in a few words, so we must refer the reader to the literature [62,63]. Nevertheless, the results of these computations are very encouraging: they show an inter-quark potential increasing with the distance stronger than linearly, i.e., at this level we can conclude that a quark/anti-quark pair must be permanently confined! For this result to appear, it is decisive that L_{eff} possess a minimum away from the origin. Due to the argument of Adler concerning the second domain where perturbative calculations are trustworthy, this minimum is already a consequence of (9.27). This is in contrast to the situation in QED, where, as we saw in section (5), L_{eff} is a monotonic function of the fields with a unique minimum at E = B = 0. The fact that in QCD the energy density is minimalized for a non-vanishing field (this is formally expressed by a non-vanishing "gluon-condensate" $\langle 0|F^a_{\mu\nu}\ F^a_{\mu\nu}\ |0\rangle \neq 0$) gave rise to the so-called Copenhagen vacuum which describes a sort of domain structure (the domains being uniformly magnetized) similar to a ferromagnetic medium. For a detailed discussion, see [64].

Another consequence of the altered sign in the β-function is that the QCD vacuum behaves as a linear, but spatially non-local medium at short distances, while at large distances it behaves as a local, but non-linear medium [59]. This is just opposite to the vacuum of electrodynamics.

With these remarks we close our outlook on the effective Lagrangian approach to quantum chromodynamics, which, in certain cases, could well be an alternative to the commonly used Monte Carlo simulations. However, the complexities of QCD once more suggest that it is preferable to first study a relatively well-understood theory, such as electrodynamics, within this context before turning to problems like the vacuum structure of QCD and quark confinement.

Appendix A: Units, Metric, Gamma Matrices

We use exclusively the Heaviside-Lorentz System with natural units, i.e., we set h = c = 1. Then the fine structure reads $\alpha = e^2/(4\pi)$.

We have, as metric

$$(g_{\mu\nu}) = (g^{\mu\nu}) = \begin{pmatrix} -1 & & & 0 \\ & +1 & & \\ & & +1 & \\ 0 & & & +1 \end{pmatrix}$$

so that the field strength tensor can be written as

$$(F^{\mu\nu}) = \begin{pmatrix} 0 & E_1 & E_2 & E_3 \\ -E_1 & 0 & B_3 & -B_2 \\ -E_2 & -B_3 & 0 & B_1 \\ -E_3 & B_2 & -B_1 & 0 \end{pmatrix}$$

The γ-matrices, which satisfy by definition the anti-commutation relations

$$\{\gamma^\mu, \gamma^\nu\} = -2g^{\mu\nu}$$

we use in the standard representation [37]

$$\gamma^0 = \begin{pmatrix} 1 & 0 \\ 0 & -1 \end{pmatrix} \quad , \quad \{\gamma^i\} \equiv \vec{\gamma} = \begin{pmatrix} 0 & \vec{\sigma} \\ -\vec{\sigma} & 0 \end{pmatrix}$$

with the Pauli matrices

$$\sigma^1 = \begin{pmatrix} 0 & 1 \\ 1 & 0 \end{pmatrix} \; , \quad \sigma^2 = \begin{pmatrix} 0 & -i \\ i & 0 \end{pmatrix} \; , \quad \sigma^3 = \begin{pmatrix} 1 & 0 \\ 0 & -1 \end{pmatrix}$$

If we now define

$$\sigma^{\mu\nu} := \frac{i}{2}[\gamma^\mu, \gamma^\nu]$$

then

$$\sigma^{ij} = \begin{pmatrix} \sigma^k & 0 \\ 0 & \sigma^k \end{pmatrix} \qquad (i,j,k = 1,2,3 \text{ cyclic}),$$

$$\sigma^{0i} = i\begin{pmatrix} 0 & \sigma^i \\ \sigma^i & 0 \end{pmatrix} \qquad (i = 1,2,3).$$

Of special importance is

$$\sigma^{12} = \begin{pmatrix} \sigma^3 & 0 \\ 0 & \sigma^3 \end{pmatrix} = \begin{pmatrix} 1 & & & \\ & -1 & & \\ & & 1 & \\ & & & -1 \end{pmatrix}$$

which is often again designated by σ^3.

We write the frequently appearing scalar product of the γ matrices with a four-vector a as

$$\not{a} = \gamma_\mu a^\mu = -\gamma^0 a^0 + \vec{\gamma}\cdot\vec{a}$$

For arbitrary four-vectors a_i, the identities apply

$$tr(\not{a}_1 \not{a}_2) = -4 a_1 \cdot a_2$$

$$tr(\not{a}_1 \not{a}_2 \not{a}_3 \not{a}_4) = 4\left[(a_1\cdot a_2)(a_3\cdot a_4) - (a_1\cdot a_3)(a_2\cdot a_4) + (a_1\cdot a_4)(a_2\cdot a_3)\right]$$

If one chooses in particular vectors with only one non-vanishing component, then it follows from the above equations that

$$tr\{\gamma_\mu \gamma_\nu\} = -4 g_{\mu\nu}$$

$$tr\{\gamma_\alpha \gamma_\beta \gamma_\delta \gamma_\varepsilon\} = 4\left[g_{\alpha\beta} g_{\delta\varepsilon} - g_{\alpha\delta} g_{\beta\varepsilon} + g_{\alpha\varepsilon} g_{\beta\delta}\right]$$

furthermore, that

$$\gamma^\mu \gamma_\mu = -4$$

$$\gamma_\mu \not{a} \gamma^\mu = 2 \not{a}$$

$$\gamma_\mu \not{a} \not{b} \gamma^\mu = 4\, a \cdot b$$

$$\gamma_\mu \not{a} \not{b} \not{c} \gamma^\mu = 2 \not{c} \not{b} \not{a}$$

Appendix B: One-Loop Effective Lagrangian of Scalar QED

In this section, with the help of a path integral [24-28], we shall derive the equation analogous to (6.1) of scalar electrodynamics and then evaluate it by means of a Zeta-function. The one-loop approximation thereby appears as the lowest correction of the classical Lagrangian in an expansion with respect to powers of ħ, which is equivalent to an expansion with respect to the number of loops. (A simple proof of this is given, for example, in [28].)

Scalar QED describes charged, spinless particles with mass m associated with the complex scalar field ϕ which interact with the electromagnetic field A^{μ}. So the classical Lagrangian is

$$\mathcal{L}(\phi,\phi^*,A) = -(\partial_\mu + ieA_\mu)\phi^*(\partial^\mu - ieA^\mu)\phi - m^2\phi^*\phi - \frac{1}{4}F_{\mu\nu}F^{\mu\nu} \qquad (B.1)$$

in order to quantize this system, first we couple the fields ϕ,ϕ^*, A to external sources η η^*, J:

$$\mathcal{L}_Q(\phi,\phi^*,A;\eta,\eta^*,J) = -(\partial_\mu + ieA_\mu)\phi^*(\partial^\mu - ieA^\mu)\phi - m^2\phi^*\phi \qquad (B.2)$$

$$-\frac{1}{4}F_{\mu\nu}F^{\mu\nu} - \frac{1}{2\xi}(\partial_\mu A^\mu)^2 + \eta^*\phi + \phi^*\eta + J_\mu A^\mu$$

In addition, a gauge-fixing term was added here to *L*. The corresponding action is defined by

$$S[\phi,\phi^*,A;\eta,\eta^*,J] = \int d^4x\, \mathcal{L}_Q(\phi,\phi^*,A;\eta,\eta^*,J)$$

The classical solutions ϕ_o, ϕ_o^*, A_o are those fields for which S becomes stationary :

$$\frac{\delta S}{\delta \phi}[\phi_0, \phi_0^*, A_0; \eta, \eta^*, J] = 0$$

$$\frac{\delta S}{\delta \phi^*}[\phi_0, \phi_0^*, A_0; \eta, \eta^*, J] = 0$$

$$\frac{\delta S}{\delta A^\mu}[\phi_0, \phi_0^*, A_0; \eta, \eta^*, J] = 0$$

One sees that ϕ_0, ϕ_0^* and A_0 are functionally dependent on the sources η, η^* and J; symbolically

$$(\phi_0, \phi_0^*, A_0) = f(\eta, \eta^*, J)$$

If we now set

$$Z[\eta, \eta^*, J] := e^{\frac{i}{\hbar}W[\eta, \eta^*, J]} := \int [d\phi\, d\phi^*\, dA]\, e^{\frac{i}{\hbar}S[\phi, \phi^*, A; \eta, \eta^*, J]} \tag{B.3}$$

then we can show [25] that Z is the generating functional of the Green's functions of the theory, while W is the generating functional of the 'connected' Green's functions; this means that the n-point functions in question are obtained by functional differentiation of Z or W followed by setting of the sources equal to zero. (In order for the path integral in (B.3) to converge, m^2 must be replaced by $m^2 - i\varepsilon$).

From W the generating functional Γ (effective action) of the one-particle irreducible, amputated (1PI) Green's function can be obtained by means of a Legendre transformation. To this end, we define the classical fields

$$\phi_c(x) := \frac{\delta W[\eta, \eta^*, J]}{\delta \eta^*(x)}$$

$$\phi_c^*(x) := \frac{\delta W[\eta, \eta^*, J]}{\delta \eta(x)}$$

$$A_c^{\mu}(x) = \frac{\delta W[\eta,\eta^*,J]}{\delta J_{\mu}(x)} \tag{B.4}$$

which are apparently functionally dependent on η, η^* and J. If one imagines this relation between (ϕ_c, ϕ_c^*, A_c) and (η, η^*, J) inverted, then it holds for Γ [25,27]

$$\Gamma[\phi_c, \phi_c^*, A_c] = W[(\eta,\eta^*,J)(\phi_c, \phi_c^*, A_c)] -$$
$$- \int d^4x \{\phi_c^* \eta(\phi_c, \phi_c^*, A_c) + \phi_c \eta^*(\phi_c, \phi_c^*, A_c) + A_c^{\mu} J_{\mu}(\phi_c, \phi_c^*, A_c)\} \tag{B.5}$$

By differentiation of Γ we get the sources back

$$\frac{\delta \Gamma[\phi_c, \phi_c^*, A_c]}{\delta \phi_c(x)} = \int d^4y \left\{ \frac{\delta W[\ldots]}{\delta \eta(y)} \frac{\delta \eta(y)}{\delta \phi_c(x)} + \frac{\delta W[\ldots]}{\delta \eta^*(y)} \frac{\delta \eta^*(y)}{\delta \phi_c(x)} + \frac{\delta W[\ldots]}{\delta J^{\mu}(y)} \frac{\delta J^{\mu}(y)}{\delta \phi_c(x)} \right\}$$

$$- \eta^*(x) \quad - \int d^4y \left\{ \phi_c^*(y) \frac{\delta \eta(y)}{\delta \phi_c(x)} + \phi_c(y) \frac{\delta \eta^*(y)}{\delta \phi_c(x)} + A_c^{\mu}(y) \frac{\delta J_{\mu}(y)}{\delta \phi_c(x)} \right\}$$

$$\underset{(B.4)}{=} - \eta^*(x)$$

With analogous formulae for the derivation with respect to ϕ_c^* and A_c, we then get all together

$$\frac{\delta \Gamma[\phi_c, \phi_c^*, A_c]}{\delta \phi_c(x)} = - \eta^*(x)$$
$$\frac{\delta \Gamma[\phi_c, \phi_c^*, A_c]}{\delta \phi_c^*(x)} = - \eta(x) \tag{B.6}$$
$$\frac{\delta \Gamma[\phi_c, \phi_c^*, A_c]}{\delta A_c^{\mu}(x)} = - J_{\mu}(x)$$

The meaning of the classical fields is clarified by

$$\phi_c(x) = \frac{\langle 0+|\phi(x)|0-\rangle^{\eta,\eta^*,J}}{\langle 0+|0-\rangle^{\eta,\eta^*,J}}$$

$$\phi_c^*(x) = \frac{\langle 0+|\phi^*(x)|0-\rangle^{\eta,\eta^*,J}}{\langle 0+|0-\rangle^{\eta,\eta^*,J}} \tag{B.7}$$

$$A_c^\mu(x) = \frac{\langle 0+|A^\mu(x)|0-\rangle^{\eta,\eta^*,J}}{\langle 0+|0-\rangle^{\eta,\eta^*,J}}$$

where we put (B.3) into (B.4) (compare [25]).

Path integrals like those in (B.3) cannot be precisely calculated for interacting theories; we must, therefore, resort to approximation methods. One possibility is offered by the saddle point method in which one first expands S around a stationary point (ϕ_o,ϕ_o^*,A_o), and terminates it after the quadratic term

$$S[\phi,\phi^*,A] = S[\phi_o,\phi_o^*,A_o] + \int d^4x \left\{\frac{\delta S[0]}{\delta\phi}(\phi-\phi_o) + \frac{\delta S[0]}{\delta\phi^*}(\phi^*-\phi_o^*) + \frac{\delta S[0]}{\delta A_\mu}(A-A_o)_\mu\right\} + S_2[\phi,\phi^*,A;\phi_o,\phi_o^*,A_o] + \dots$$

Here, $[0] \equiv [\phi_o,\phi_o^*,A_o]$ and

$$\frac{\delta S[0]}{\delta\phi}(\phi-\phi_o) = \frac{\delta S[0]}{\delta\phi(x)}(\phi(x)-\phi_o(x))$$

were introduced for the sake of shortness; furthermore, S_2 denotes the quadratic term of the Taylor series which we shall write out later. Since the classical action S for (ϕ_o,ϕ_o^*,A_o) is required to be stationary, the linear term of this expansion vanishes so that it follows from (B.3) in the approximation of the saddle point method

$$e^{\frac{i}{\hbar}W[\eta,\eta^*,J]} = \exp\left\{\frac{i}{\hbar}S[(\phi_0,\phi_0^*,A_0)(\eta,\eta^*,J);\eta,\eta^*,J]\right\}\cdot$$

$$\cdot\int[d\phi\, d\phi^*\, dA]\exp\left\{\frac{i}{\hbar}S_2[\phi,\phi^*,A;(\phi_0,\phi_0^*,A_0)(\eta,\eta^*,J)]\right\}$$

This particular notation signifies that ϕ_o, ϕ_o^* and A_o are dependent on the sources η, η^* and J. So,

$$W[\eta,\eta^*,J] = S[(\phi_0,\phi_0^*,A_0)(\eta,\eta^*,J);\eta,\eta^*,J] + \hbar W_1[(\phi_0,\phi_0^*,A_0)(\eta,\eta^*,J)] + O(\hbar^2) \tag{B.8}$$

with

$$W_1[(\phi_0,\phi_0^*,A_0)(\eta,\eta^*,J)] := -i\ln\int[d\phi\, d\phi^*\, dA]\cdot \exp\left\{\frac{i}{\hbar}S_2[\phi,\phi^*,A;(\phi_0,\phi_0^*,A_0)(\eta,\eta^*,J)]\right\} \tag{B.9}$$

Before further evaluating (B.9), let us consider the so-called tree approximation of (B.8) in which we neglect the term of the order $\hbar$ and simply set

$$W[\eta,\eta^*,J] = S[(\phi_0,\phi_0^*,A_0)(\eta,\eta^*,J);\eta,\eta^*,J] + O(\hbar)$$

From (B4), it follows then that

$$\phi_c(x) \equiv \frac{\delta W[\eta,\eta^*,J]}{\delta\eta^*(x)} = \frac{\delta S[(\phi_0,\phi_0^*,A_0)(\eta,\eta^*,J);\eta,\eta^*,J]}{\delta\eta^*(x)} + O(\hbar)$$

$$= \phi_0(x) + \int d^4y\left\{\underbrace{\frac{\delta S[\cdots]}{\delta\phi_0(y)}}_{=0}\frac{\delta\phi_0(y)}{\delta\eta^*(x)} + \underbrace{\frac{\delta S[\cdots]}{\delta\phi_0^*(y)}}_{=0}\frac{\delta\phi_0^*(y)}{\delta\eta^*(x)} + \underbrace{\frac{\delta S[\cdots]}{\delta A_0^*(y)}}_{=0}\frac{\delta A_0(y)}{\delta\eta^*(x)}\right\} + O(\hbar)$$

$$= \phi_0(x) + O(\hbar)$$

$$\phi_c^*(x) = \phi_0^*(x) + O(\hbar)$$

$$A_c(x) = A_0(x) + O(\hbar)$$

The fields (B.4) in this approximation are, then, identical with the solutions of the classical equations of motion. For the effective action Γ it follows that

$$\Gamma[\phi_c,\phi_c^*,A_c] = S[\phi_c,\phi_c^*,A_c]$$

$$- \int d^4x \{\phi_c^*\eta(\phi_c,\phi_c^*,A_c) + \phi_c\eta^*(\phi_c,\phi_c^*,A_c)$$

$$+ A_c^\mu J_\mu(\phi_c,\phi_c^*,A_c)\} + O(\hbar)$$

$$= \int d^4x \{\mathcal{L}(\phi_c,\phi_c^*,A_c) - \frac{1}{2\xi}(\partial_\mu A_c^\mu)^2\} + O(\hbar)$$

So the 1PI-Green's functions in the tree approximation are generated by the classical action!

We now turn back to (B.8) and calculate the term proportional to ħ (one-loop approximation); for this, the following definitions are useful

$$\phi_c =: \phi_0 + \hat{\phi} + O(\hbar^2) \quad , \quad \hat{\phi} = \hbar \frac{\delta W_1}{\delta \eta^*}$$

$$\phi_c^* =: \phi_0^* + \hat{\phi}^* + O(\hbar^2) \quad , \quad \hat{\phi}^* = \hbar \frac{\delta W_1}{\delta \eta}$$

$$A_c^\mu =: A_0^\mu + \hat{A}^\mu + O(\hbar^2) \quad , \quad \hat{A}^\mu = \hbar \frac{\delta W_1}{\delta J_\mu}$$

Since (ϕ_0,ϕ_0^*,A_0) satisfies the classical equations of motion, we have

$$S[\phi_c,\phi_c^*,A_c] = S[\phi_0+\hat{\phi},\phi_0^*+\hat{\phi}^*,A_0+\hat{A}] + O(\hbar^2)$$

$$= S[\phi_0,\phi_0^*,A_0] + \int d^4x \left\{\frac{\delta S[0]}{\delta\phi_0}\hat{\phi} + \frac{\delta S[0]}{\delta\phi_0^*}\hat{\phi}^* + \frac{\delta S[0]}{\delta A_0}\hat{A}\right\} + O(\hbar^2)$$

$$= S[\phi_0,\phi_0^*,A_0] + O(\hbar^2)$$

Furthermore,

$$\hbar W_1[\phi_c, \phi_c^*, A_c] = \hbar W_1 [\phi_0, \phi_0^*, A_0] + O(\hbar^2)$$

So, for the effective action Γ it follows that

$$\Gamma[\phi_c, \phi_c^*, A_c] = S[\phi_c, \phi_c^*, A_c; (\eta, \eta^*, J)(\phi_c, \phi_c^*, A_c)] + \hbar W_1 [\phi_0, \phi_0^*, A_0]$$

$$- \int d^4x \{\phi_c \eta^*(\phi_c, \phi_c^*, A_c) + \phi_c^* \eta(\phi_c, \phi_c^*, A_c) + A_c^\mu J_\mu(\phi_c, \phi_c^*, A_c)\} + O(\hbar^2)$$

$$= \int d^4x \{\mathcal{L}(\phi_c, \phi_c^*, A_c) - \frac{1}{2\xi}(\partial_\mu A_c^\mu)^2\} + \hbar W_1 [\phi_c, \phi_c^*, A_c] + O(\hbar^2)$$

The one-loop correction Γ_1 to the effective action is then

$$\Gamma_1[\phi_c, \phi_c^*, A_c] = \hbar W_1 [\phi_c, \phi_c^*, A_c]$$

$$= -i\hbar \ln \int [d\phi d\phi^* dA] \exp\{\frac{i}{\hbar} S_2[\phi, \phi^*, A; \phi_c, \phi_c^*, A_c]\} \tag{B.10}$$

We shall evaluate the path integral only for the special case

$$\phi_c = \phi_c^* = 0 \quad , \quad A_c^\mu(x) = -\frac{1}{2} F^{\mu\nu} x_\nu , \quad F_{12} = -F_{21} = B \quad ;$$

$$= Const$$

if we define the one-loop effective potential V_1 by

$$\Gamma_1 [0, 0, A_c] =: - \int d^4x \, V_1(B)$$

then

$$V_1(B) = \left(\int d^4x\right)^{-1} i\hbar \ln \int [d\phi d\phi^* dA] \exp\{\frac{i}{\hbar} S_2[\phi, \phi^*, A; 0, 0, A_c]\} \tag{B.11}$$

Here, the functional S_2 is

$$S_2 [\phi, \phi^*, A; 0, 0, A_c] = \frac{1}{2} \int dx dy \{(A - A_c) \frac{\delta^2 S[0]}{\delta A \, \delta A} (A - A_c) +$$

$$+\phi\frac{\delta^2 S[0]}{\delta\phi\delta\phi}\phi+\phi^*\frac{\delta^2 S[0]}{\delta\phi^*\delta\phi^*}\phi^*+2\phi^*\frac{\delta^2 S[0]}{\delta\phi^*\delta\phi}\phi$$

$$+2\phi\frac{\delta^2 S[0]}{\delta\phi\delta A}(A-A_c)+2\phi^*\frac{\delta^2 S[0]}{\delta\phi^*\delta A}(A-A_c)\Big\}$$

$$=\frac{1}{2}\int d^4x\,d^4y\,\Big\{(A-A_c)\frac{\delta^2 S[0]}{\delta A\delta A}(A-A_c)+2\phi^*\frac{\delta^2 S[0]}{\delta\phi^*\delta\phi}\phi\Big\} \qquad \text{(B.12)}$$

with

$$(A-A_c)\frac{\delta^2 S[0]}{\delta A\delta A}(A-A_c)\equiv(A^\mu(x)-A_c^\mu(x))\frac{\delta^2 S[0]}{\delta A^\mu(x)\delta A^\nu(y)}(A^\nu(y)-A_c^\nu(y))$$

After a shift of the variables of integration ($A\rightarrow A+A_c$), (B.11) and (B.12) show that the path integral over A changes the effective potential V_1 by only a constant which we can ignore. Thus, it follows from (B.11) after a Wick rotation $t\rightarrow\tau=it$ and with $\hbar=1$ that

$$V_1(B)=-\Omega^{-1}\ln\int[d\phi^*d\phi]\,e^{-\int\phi^*M_E\phi} \qquad \text{(B.13)}$$

with

$$\Omega\equiv\int d^3x\,d\tau=i\int d^4x$$

and

$$M_E\equiv-\left(\frac{\delta^2 S[0]}{\delta\phi^*\delta\phi}\right)_E$$

$$=m^2-(\partial-ieA_c)^2_E$$

$$\equiv m^2+\pi^2_E$$

The Gaussian path integral in (B.13) can be easily evaluated [12] and gives

$$V_1(B) = -\Omega^{-1} \ln(\det M_E)^{-1}$$

$$= \Omega^{-1} \ln \det M_E$$

In order to make this equation right, also with respect to the dimensions, we again introduce an at first arbitrary factor μ^2

$$V_1(B) = \Omega^{-1} \ln \det(M_E/\mu^2)$$

With (6.4) it follows that

$$\begin{aligned} V_1(B) &= \Omega^{-1} \ln \exp\left[-\zeta'_{M_E/\mu^2}(0)\right] \\ &= -\Omega^{-1} \zeta'_{M_E/\mu^2}(0) \\ &=: -\Omega^{-1} \zeta'_1(0) \end{aligned} \tag{B.14}$$

According to (6.17), the spectrum of M_E/μ^2 is

$$\left\{[m^2+K_0^2+K_3^2+(2n+1)eB]\mu^{-2} \,\middle|\, K_0, K_3 \in \mathbb{R}, n \in \mathbb{N}\right\}$$

so that, in analogy to (6.18) it follows for the Zeta-function that

$$\zeta_1(s) = \mu^{2s} \Omega \sum_{n=0}^{\infty} \frac{eB}{2\pi} \iint_{-\infty}^{\infty} \frac{dK_0 dK_3}{(2\pi)^2} \left[m^2+K_0^2+K_3^2+(2n+1)eB\right]^{-s}$$

From this, with the integral (6.20) and (6.21), we get

$$\zeta_1(s) = \mu^{2s} \Omega \sum_{n=0}^{\infty} \frac{eB}{(2\pi)^3} B\left(\tfrac{1}{2}, s-\tfrac{1}{2}\right) \int_{-\infty}^{\infty} dK_3 \left[m^2+K_3^2+(2n+1)eB\right]^{\frac{1}{2}-s}$$

$$= \mu^{2s} \Omega \frac{eB}{8\pi^2} (s-1)^{-1} (2eB)^{1-s} \sum_{n=0}^{\infty} \left[\frac{m^2+eB}{2eB} + n\right]^{1-s}$$

$$= \mu^{2s} \frac{\Omega}{16\pi^2} (2eB)^{2-s} (s-1)^{-1} \zeta(s-1,q) , \quad q := \frac{m^2+eB}{2eB}$$

The derivative results in

$$\zeta_1'(s) = \frac{\Omega}{16\pi^2} \Big[\ln\mu^2 \, \mu^{2s} (2eB)^{2-s} (s-1)^{-1} \zeta(s-1,q)$$

$$- \mu^{2s} \ln(2eB) (2eB)^{2-s} (s-1)^{-1} \zeta(s-1,q)$$

$$- \mu^{2s} (2eB)^{2-s} (s-1)^{-2} \zeta(s-1,q)$$

$$+ \mu^{2s} (2eB)^{2-s} (s-1)^{-1} \zeta'(s-1,q) \Big]$$

So, for s = 0,

$$\zeta_1'(0) = \Omega \frac{(eB)^2}{8\pi^2} \Big[-2\zeta(-1,q) \{ 1 - \ln \frac{2eB}{\mu^2} \} - 2\zeta'(-1,q) \Big]$$

Because of (6.30), we get

$$-2\zeta(-1,q) = q^2 - q + \frac{1}{6}$$

$$= \left(\frac{m^2}{2eB}\right)^2 - \frac{1}{12}$$

and thus

$$\zeta_1'(0) = \Omega \frac{(eB)^2}{8\pi^2} \Big\{ \Big[\Big(\frac{m^2}{2eB}\Big)^2 - \frac{1}{12} \Big] \Big(1 - \ln \frac{2eB}{\mu^2}\Big) - 2\zeta'(-1,q) \Big\}$$

In accordance with (B.14), the one-loop effective Lagrangian is then given by

$$\mathcal{L}^{(1)}(B) = - V_1(B) \tag{B.15}$$

$$= \Omega^{-1} \zeta_1'(0)$$

$$= \frac{(eB)^2}{8\pi^2} \Big\{ \Big[\Big(\frac{m^2}{2eB}\Big)^2 - \frac{1}{12} \Big] \Big(1 - \ln \frac{2eB}{m^2}\Big) - 2\zeta'(-1,q) \Big\}$$

$$= \frac{1}{16\pi^2} \Big\{ \frac{1}{2} \Big[m^4 - \frac{1}{3}(eB)^2 \Big] \Big(1 - \ln \frac{2eB}{m^2}\Big)$$

$$- 4(eB)^2 \zeta'\Big(-1, \frac{m^2+eB}{2eB}\Big) \Big\}$$

Here, we have again chosen $\mu = m$ as mass scale. With the exception of an unimportant constant, (B.15) can also be written as

$$\mathcal{L}^{(1)}(B) = \frac{1}{64\pi^2}\{[2m^4 - \tfrac{2}{3}(eB)^2]\left(1 + \ln\frac{m^2}{2eB}\right) - 3m^4 - (4eB)^2\,\zeta'\left(-1, \frac{m^2+eB}{2eB}\right)$$

this is exactly the result obtained by Dittrich [8] using a dimensional regularization!

Appendix C: The Casimir Effect

In this appendix, the analogy mentioned in section 6 between $L^{(1)}$ at finite temperature and the Casimir effect will be more closely investigated.

The Casimir effect is a non-classical electromagnetic, attractive or repulsive force which occurs between electrically neutral conductors in a vacuum. The size of this force was first calculated by Casimir [20] for the case of ideal conducting, infinitely extended, parallel plates; his result was a force

$$F = -\frac{\pi^2}{240} \cdot \frac{\hbar c}{a^4} \tag{C.1}$$

where a is the distance between the plates and the negative sign indicates that the plates attract each other. This force apparently depends only on the fundamental constants $\hbar$ and c apart from the distance between the plates; not, however, of the coupling constant α between the Maxwell and the matter field. Its quantum mechanical character is revealed by the fact that F vanishes in the classical limit $\hbar \to 0$.

Casimir's derivation of (C.1) was based on the concept of a quantum electrodynamic (particle) vacuum representing the zero-point oscillations of an infinite number of harmonic oscillators. As a result, one gets the total vacuum energy by summation over the zero-point energies $\frac{1}{2}\hbar\omega_{\vec{K}}$ of all allowed modes with wave number vector $\vec{k}$ and polarization σ

$$E = \sum_{\vec{k},\sigma} \tfrac{1}{2}\hbar\omega_{\vec{k}} \tag{C.2}$$

If we evaluate (C.2) for the case of two plane parallel plates at distance a from each other, one does get a divergent total energy E(a), but the energy difference E(a)-E(a+δa) is finite (δa = infinitesimal change in the plate distance), leading also to a finite force per unit area

$$F = -\frac{\partial E(a)}{\partial a} \tag{C.3}$$

To calculate this energy difference or force, a UV-cut-off is usually introduced, i.e., (C.2) is replaced by

$$\sum_{\vec{k},\sigma} \tfrac{1}{2}\hbar\omega_{\vec{k}}\, e^{-\frac{b}{\pi c}\omega_{\vec{k}}}$$

and, in the end result, the limit $b \to 0$ is considered.

This derivation of (C.1), however, can give the impression that the appearance of the Casimir force is linked to the existence of the zero-point fluctuations of the quantized electromagnetic field. Since it has been speculated that the real Hamiltonian of the field [21] could not contain the term (C.2), this would mean that the Casimir force would also not appear. That these assumptions do not apply was shown by Schwinger [22] in the framework of Source Theory [29], which does totally without the concept of zero-point fluctuations of the field, i.e., a structured vacuum. Besides, the Casimir effect was proven experimentally [23].

In the following, we shall consider the problem according to Hawking [15] from the viewpoint of path integral quantization and Zeta-function regularization. Here, it is again unnecessary to refer to the vacuum oscillation. For reasons of simplicity, we wish to consider the Casimir effect only for a real, scalar field theory which is defined by ($\hbar = c = 1$!)

$$\mathcal{L}(\phi) = -\frac{1}{2}\partial_\mu\phi\,\partial^\mu\phi - \frac{1}{2}m^2\phi^2 - V(\phi) \tag{C.4}$$

with the arbitrary potential V.
In order to carry over the result obtained into QED, we shall only have to multiply F by a factor 2 (corresponding to the two polarization states of the photon). First, we couple the field ϕ to an external source J

$$\mathcal{L}(\phi) \longrightarrow \mathcal{L}(\phi) + J\phi \tag{C.5}$$

According to [25], we can then write the vacuum amplitude $\langle 0_+|0_-\rangle^J$ or the action W[J] in the form

$$\langle 0_+|0_-\rangle^J = e^{iW[J]} = \int[d\phi]\,e^{i\int d^4x\,\{\mathcal{L}(\phi) + J\phi\}} \tag{C.6}$$

where we guarantee the convergence of the path integral by the substitution $m^2 \to m^2 - i\varepsilon$, $\varepsilon > 0$. Until now, we have assumed that $|0_-\rangle$ or $|0_+\rangle$ describes a vacuum which is not 'disturbed' by the presence of certain geometries, i.e., the path integral (C.6) is, without restriction by boundary conditions, to be taken over all fields ϕ. This changes as soon as we introduce two plates into the vacuum, for example, perpendicular to the z-axis (points of intersection: $z = 0$ and $z = a$) and require that only those fields should contribute to the path integral which vanish on the plate surfaces, i.e., for which it holds that

$$\phi(x_0, x_1, x_2, 0) = \phi(x_0, x_1, x_2, a) = 0 \tag{C.7}$$

for arbitrary (x_0, x_1, x_2). In QED such boundary conditions can be fulfilled by the use of perfect conducting surfaces. Instead of (C.6), we now get

$$\langle 0+|0-\rangle_a^J = e^{iW(a,[J])}$$
$$= \int_{F_a} [d\phi] \exp\left[i\int d^4x \left\{-\tfrac{1}{2}\partial_\mu\phi\,\partial^\mu\phi - \tfrac{1}{2}(m^2 - i\varepsilon)\phi^2 - V(\phi) - J\phi\right\}\right] \tag{C.8}$$

where $\int_{Fa}$ suggests that the path integral is only to be taken over the restricted space of functions Fa defined by (C.7). With this, we have represented the vacuum amplitude or the action for the most general case as a function of the geometric parameter a and as a functional of the external source J. In order to approach the conditions of the QED Casimir effect, we now choose J = $\underline{0}$ as well as a free (V = $\underline{0}$), masslees (m = 0) field ϕ. Following a partial integration:

$$\langle 0+|0-\rangle_a = e^{iW(a)} = \int_{F_a} [d\phi]\, e^{-\frac{i}{2}\int d^4x\, \phi\{-\partial^2 - i\varepsilon\}\phi} \tag{C.9}$$

The Gauss integral gives [12]

$$\langle 0+|0-\rangle_a = e^{iW(a)} = \int_{F_a} [d\phi]\, e^{-\frac{1}{2}\int d^3x\, d\tau\, \phi\{-\Box_E\}\phi}$$
$$= N\left[\det\left(-\Box_E / F_a\right)\right]^{-\frac{1}{2}} \tag{C.10}$$

Here, N is a (divergent) constant which we shall set = 1, since it only contributes a non-physical additive constant to W(a). By writing $\Box_E/Fa$, we mean that only eigenvalues with eigenfunctions in Fa can be used to evaluate the determinant. Furthermore (in keeping with the $i\varepsilon$ requirement), a Wick rotation $t \to i\tau$ was made, i.e., $\Box_E = \partial_\tau^2 + \Delta$.

With (6.4), it follows that

$$\langle 0+|0-\rangle_a = e^{iW(a)} = \left[\exp\{-\zeta'_{-\Box_E/\mathcal{F}_a}(0)\}\right]^{-\frac{1}{2}} \tag{C.11}$$

$$= \exp\left[\tfrac{1}{2}\zeta'_{-\Box_E/\mathcal{F}_a}(0)\right].$$

The operator $-\Box_E/Fa$ has the spectrum

$$\left\{K_0^2 + K_1^2 + K_2^2 + \left(\frac{\pi n}{a}\right)^2 \,\middle|\, K_0, K_1, K_2 \in \mathbb{R}, n \in \mathbb{N}\right\}$$

and thus, the Zeta function

$$\zeta_{-\Box_E/\mathcal{F}_a}(s) = 2\frac{A}{(2\pi)^2}\frac{T_E}{2\pi}\iiint_{-\infty}^{\infty} dK_0\,dK_1\,dK_2 \sum_{n=1}^{\infty}\left[K_0^2+K_1^2+K_2^2+\left(\frac{n\pi}{a}\right)^2\right]^{-s} \tag{C.12}$$

Here, the factor 2 makes allowance for the two polarization possiblities of the photon, which, in our simple model, have no analogue. Furthermore, AT_E is a normalization volume in three dimensional (0,1,2) space, where the Euclidean time T_E is linked to a (Minkowski) normalization time interval T by $T_E = iT$. Dropping the term independent of a (n = 0) in (C.12) simply leads to the subtraction of an (infinite) constant of W(a).

A comparison with (6.52) now shows, that $\zeta_{-\Box_E/Fa}$ differs from $\zeta_1^\beta(\zeta_2^\beta)$ only in that the discretization of the k-integration was not performed in the time component, but in the space component; furthermore, one can see that the parameter a corresponds to $\beta = 1/kT$.

Further evaluation of (C.12) now takes on exactly the same form as in section 6:

$$\zeta_{-\Box_E/\mathcal{F}_a}(s) = 2AT_E \frac{4\pi}{(2\pi)^3} \sum_{n=1}^{\infty} \int_0^{\infty} dk\; k^2 \left[k^2 + \left(\frac{n\pi}{a}\right)^2\right]^{-s}$$

$$= \frac{8\pi}{(2\pi)^3} AT_E \left(\frac{\pi}{a}\right)^{3-2s} \sum_{n=1}^{\infty} n^{3-2s} \frac{1}{2} \frac{\Gamma(\frac{3}{2})\Gamma(s-\frac{3}{2})}{\Gamma(s)}$$

$$= \frac{4\pi}{(2\pi)^3} AT_E \left(\frac{\pi}{a}\right)^{3-2s} \zeta(2s-3) \frac{\Gamma(\frac{3}{2})\Gamma(s-\frac{3}{2})}{\Gamma(s)}$$

The derivative is

$$\zeta'_{-\Box_E/\mathcal{F}_a}(0) = \frac{4\pi}{(2\pi)^3} AT_E \left(\frac{\pi}{a}\right)^3 \zeta(-3)\,\Gamma(\tfrac{3}{2})\,\Gamma(-\tfrac{3}{2}) \frac{d}{ds} \frac{1}{\Gamma(s)}\bigg|_{s=0}$$

$$= \frac{\pi^2}{360a^3} AT_E$$

From (C.11) we get

$$\langle 0+|0-\rangle = e^{iW(a)} = e^{-\varepsilon(a)T_E} = e^{-i\varepsilon(a)T}$$

with

$$\varepsilon(a) = -\frac{\pi^2}{720a^3} A$$

The appearance of the phase factor $e^{-i\varepsilon(a)T}$ in the vacuum amplitude allows us to identify $\varepsilon(a)$ as the vacuum energy displacement and to write, for the force per surface unit

$$F = -\frac{1}{A} \frac{\partial \varepsilon}{\partial a}$$

which leads to

$$\mathcal{F} = -\frac{\pi^2}{240}\cdot\frac{1}{a^4}$$

or, after putting in $\hbar$ and c

$$\mathcal{F} = -\frac{\pi^2}{240}\cdot\frac{\hbar c}{a^4} \tag{C.13}$$

This is precisely Casimir's result which we have now completely derived without the concept of a structured vacuum!

We should now like to briefly demonstrate the relation between the above-described method and Schwinger's approach [22] which, by means of a causal analysis in the framework of source theory, beginning with the vacuum amplitude in presence of an external source J

$$\langle 0_+ | 0_- \rangle^J = e^{iW[J]}$$

arrives at the following differential expression for the action:

$$\begin{aligned} \delta W &= \frac{i}{2}\int d^4x\, d^4x'\; D_a(x,x')\,\delta D_a^{-1}(x',x) \\ &\equiv \frac{i}{2}\,\mathrm{Tr}\,(D_a\,\delta D_a^{-1}) \\ &= \frac{i}{2}\,\delta\,(\mathrm{Tr}\,\ln D_a^{-1}) \end{aligned} \tag{C.14}$$

Here, 'δ' refers to variation of the plate distance, and D_a is the Green's function of the probelm, defined by

$$\Box D_a(x,x') = -\delta(x-x')\,,\quad D_a(x_0,x_1,x_2,0;x') = D_a(x_0,x_1,x_2,a;x') = 0$$

Integration of (C.14) gives, disregarding an unimportant additive constant,

$$W(a) = \frac{i}{2}\,\mathrm{Tr}\,\ln \mathcal{D}_a^{-1} \qquad \text{(C.15)}$$
$$= \frac{i}{2}\,\ln\det \mathcal{D}_a^{-1}$$

In [22] the evaluation of (C.14) or (C.15) follows in such a manner that the Green's function D_a is explicitly calculated and inserted into (C.14). If one is only interested in W(a), however, then knowledge of D_a is not necessary, since the determinant in (C.15) can be calculated by means of a Zeta-function whose construction requires only the spectrum D_{aE}^{-1} . But this agrees with that of $-\Box_E/Fa$ analogous to our calculation. This means much less calculatory expense in comparison with [22].

In concluding, we wish to mention again that one can naturally also get (C.13) from the formula for the vacuum energy often used in relation to the confinement problem (compare, for example [45,46]):

$$E = -\lim_{T_E\to\infty} \frac{1}{T_E} \ln \int [d\phi]\, e^{-S_E[\phi]}$$

(S_E = Euclidean action). In our case,

$$E = -\lim_{T_E\to\infty} \frac{1}{T_E} \ln \int_{\mathcal{F}_a} [d\phi]\, e^{-\frac{1}{2}\int d^4x_E\, \phi(-\Box_E)\phi}$$

$$= -\lim_{T_E\to\infty} \frac{1}{T_E} \ln\left[\det(-\Box_E/\mathcal{F}_a)\right]^{-\frac{1}{2}}$$

$$= -\lim_{T_E\to\infty} \frac{1}{T_E}\,\frac{1}{2}\,\zeta'_{-\Box/\mathcal{F}_a}(0)$$

$$= -\frac{1}{2}\frac{1}{T_E}\frac{\pi^2}{360a^3} A\, T_E$$

$$= -\frac{\pi^2}{720\, a^3} A$$

which again gives

$$F = -\frac{1}{A}\frac{\partial E}{\partial a} = -\frac{\pi^2}{240}\frac{\hbar c}{a^4} \quad .$$

Appendix D: Derivatives of W[A]

We now calculate the first two functional derivatives of [47]

$$iW[A] = -\mathrm{Tr}\,\ln\,(1-e\,\gamma A\;G_+)^{-1}$$
$$= +\,\mathrm{Tr}\,\ln\,(1-e\,\gamma A\;G_+) \qquad \text{(D.1)}$$

with respect to the potential A. To do so, we first write the logarithm in (D.1)

$$\ln(1-x) = -x-\tfrac{1}{2}x^2-\tfrac{1}{3}x^3 - \cdots$$
$$= -\int_0^1 d\lambda\,(x+\lambda x^2+\lambda^2x^3+\cdots)$$
$$= -\int_0^1 d\lambda\;x\,(1+\lambda x+(\lambda x)^2+\cdots)$$
$$= -\int_0^1 d\lambda\;x\,\frac{1}{1-\lambda x}\;, \qquad x \equiv e\,\gamma A\;G_+$$

in the form

$$\ln(1-e\,\gamma A\,G_+) = -\int_0^1 d\lambda\; e\,\gamma A\;G_+\,\frac{1}{1-e\,\gamma A\,G_+\lambda}$$
$$= -\int_0^e de'\;\gamma A\;G_+\,\frac{1}{1-e'\,\gamma A\,G_+}$$

where, at the end, e' was set equal to λe. So (D.1) gives

$$iW[A] = \mathrm{Tr}\,\ln\,(1-e\,\gamma A\,G_+)$$
$$= -\int_0^e de'\;\mathrm{Tr}\left[\gamma A\;G_+(1-e'\,\gamma A\,G_+)^{-1}\right]$$
$$\underset{(5.3)}{=} -\int_0^e de'\;\mathrm{Tr}\left[\gamma A\;G_+[e'A]\right] \qquad \text{(D.2)}$$
$$= -\int_0^e de'\;\mathrm{tr}\left[\int d^4x\;\gamma A(x)\;G_+(x,x|e'A)\right].$$

Now we show that, for the derivative of the propagator in an external field

$$\frac{\delta G_+(x,y|A)}{\delta A^\mu(z)} = G_+(x,z|A)\, e\gamma_\mu\, G_+(z,y|A) \qquad \text{(D.3)}$$

is valid. For this, we begin with the defining equation for $G_+[A]$

$$\left[m+\gamma\left(\tfrac{1}{i}\partial_x - eA(x)\right)\right] G_+(x,y|A) = \delta(x-y)$$

to which the integral equation

$$G_+(x,y|A) = G_+(x-y) + \int du\, G_+(x-u)\, e\gamma_\mu A^\mu(u)\, G_+(u,y|A)$$

is equivalent. Taking the derivative results in

$$\begin{aligned}\frac{\delta G_+(x,y|A)}{\delta A^\mu(z)} &= \int du\Big[G_+(x-u)\, e\gamma_\mu\, \delta(u-z)\, G_+(u,y|A) \\ &\qquad + G_+(x-u)\, e\gamma^\nu A_\nu(u) \frac{\delta G_+(u,y|A)}{\delta A^\mu(z)}\Big] \\ &= G_+(x-z)\, e\gamma_\mu\, G_+(z,y|A) \\ &\qquad + \int du\, G_+(x-u)\, e\gamma^\nu A_\nu(u) \frac{\delta G_+(u,y|A)}{\delta A^\mu(z)}\end{aligned}$$

It then follows that

$$\int du\left[\delta(x-u) - G_+(x-u)\, e\gamma A(u)\right]\frac{\delta G_+(u,y|A)}{\delta A^\mu(z)} = G_+(x-z)\, e\gamma_\mu\, G_+(z,y|A)$$

written as a matrix, this equation reads

$$\int du\, \langle x|1-G_+ e\gamma A|u\rangle\langle u|\frac{\delta G_+[A]}{\delta A^\mu(z)}|y\rangle = \langle x|G_+|z\rangle\, e\gamma_\mu \langle z|G_+[A]|y\rangle$$

or simply

$$(1-G_+ e\gamma A)\frac{\delta G_+[A]}{\delta A^\mu} = G_+\, e\gamma_\mu\, G_+[A]$$

which immediately leads to

$$\frac{\delta G_+[A]}{\delta A^\mu} = (1 - G_+ e\gamma A)^{-1}\, G_+\, e\gamma_\mu\, G_+[A]$$

$$= G_+[A]\, e\gamma_\mu\, G_+[A]$$

If we now return to space representation, then (D.3) follows exactly, qed. So, for the derivative of (D.2) we get

$$\frac{i\,\delta W[A]}{\delta A^\mu(u)} = -\, tr \int_0^e de' \int dx \frac{\delta}{\delta A^\mu(u)} [\gamma_\alpha A^\alpha(x)\, G_+(x,x|e'A)]$$

$$= -\, tr \int_0^e de' \int dx\, [\gamma_\mu\, \delta(x-u)\, G_+(x,x|e'A) + \gamma_\alpha A^\alpha(x) \frac{\delta G_+(x,x|e'A)}{\delta A^\mu(u)}]$$

$$= -\, tr \int_0^e de' \int dx\, [\gamma_\mu\, \delta(x-u)\, G_+(x,x|e'A) + \gamma A(x)\, G_+(x,u|e'A)\, e'\gamma_\mu\, G(u,x|e'A)]$$

$$= -\int_0^e de' \,[tr\{\gamma_\mu G_+(u,u|e'A)\} + tr \int dx\, G_+(u,x|e'A)\gamma A(x)\, G_+(x,u|e'A)\, e'\gamma_\mu]$$

$$\underset{(*)}{=} -\int_0^e de' [tr\{\gamma_\mu G_+(u,u|e'A)\} + tr\{\frac{\partial}{\partial e'}(G_+(u,u|e'A))\, e'\gamma_\mu\}]$$

$$= -\int_0^e de' \frac{\partial}{\partial e'} [tr\{e'\gamma_\mu G_+(u,u|e'A)\}]$$

$$= -e\, tr\,[\gamma_\mu G_+(u,u|eA)] \tag{D.4}$$

In the fourth line, the cyclicity of the trace was utilized, and in (*) we made use of

$$\frac{\partial}{\partial e'} G_+[e'A] = \frac{\partial}{\partial e'} G_+ (1 - e'\gamma A\, G_+)^{-1}$$

$$= G_+ (1 - e' \gamma A G_+)^{-1} \gamma A G_+ (1 - e' \gamma A G_+)^{-1}$$

$$= G_+ [e'A] \gamma A \; G_+ [e'A]$$

i.e., of

$$\int dx \, G_+(u, x | e'A) \, \gamma A(x) \, G_+(x, u | e'A) = \frac{\partial}{\partial e'} G_+(u, u | e'A)$$

From (D.4), it follows for the second derivative:

$$i \frac{\delta^2 W[A]}{\delta A^{\mu_1}(x_1) \, \delta A^{\mu_2}(x_2)} = \frac{\delta}{\delta A^{\mu_1}(x_1)} (-e) \, tr \left[\gamma_{\mu_2} G_+(x_2, x_2 | A) \right]$$

$$= -e \; tr \left[\gamma_{\mu_2} \frac{\delta G_+(x_2, x_2 | A)}{\delta A^{\mu_1}(x_1)} \right]$$

$$= -e^2 \, tr \left[\gamma_{\mu_2} G_+(x_2, x_1 | A) \gamma_{\mu_1} G_+(x_1, x_2 | A) \right]$$

$$= -e^2 \, tr \left[\gamma_{\mu_1} G_+(x_1, x_2 | A) \, \gamma_{\mu_2} G_+(x_2, x_1 | A) \right]$$

So, in summary, we have found

$$i \frac{\delta W[A]}{\delta A^{\mu}(x)} = - e \, tr \left[\gamma_\mu G_+(x, x | A) \right]$$

$$i \frac{\delta^2 W[A]}{\delta A^{\mu_1}(x_1) \, \delta A^{\mu_2}(x_2)} = - e^2 \, tr \left[\gamma_{\mu_1} G_+(x_1, x_2 | A) \, \gamma_{\mu_2} G_+(x_2, x_1 | A) \right] \tag{D.5}$$

Appendix E: Power Series and Laurent Series of K(z,v)

In this appendix we calculate the power series of K = K(z) to the quadratic term, necessary for the renormalization of $L^{(2)}$, and the singular part of the Laurent series of $K = K(v^2)$ about $v^2 = 1$. Here, K(z,v) is given by (7.19) with (7.21).

We first turn our attention to the power series of

$$K_1 = \underbrace{\frac{16}{\sin z}}_{=:C} \Bigg[\underbrace{\frac{z^2 \sin z\,(\cos zv - v\cot z \cdot \sin zv)}{(1-v^2)\,(\cos zv - \cos z)}}_{=:A} \qquad \text{(E.1)}$$

$$+ \underbrace{\frac{z^2\,(\cos z \cdot \sin z - z)}{2(\cos zv - \cos z) - (1-v^2)\, z \sin z}}_{=:B} \Bigg]$$

Then

$$A \equiv \frac{\sin z}{1-v^2} \underbrace{\left[\cos zv - v\cot z \cdot \sin zv\right]}_{=:D_1} \underbrace{\frac{2z^2}{2(\cos zv - \cos z)}}_{=:D_2}$$

Expansion of D_1 and D_2 gives

$$D_1 = \cos zv - v\cot z \cdot \sin zv$$

$$= \left(1 - \frac{z^2v^2}{2!} + \frac{z^4v^4}{4!} + \dots\right) - v\left(\frac{1}{z} - \frac{1}{3}z - \frac{z^3}{45} + \dots\right)\Big(zv -$$

$$- \frac{1}{6}z^3v^3 + \frac{1}{5!}z^5v^5 + \dots\Big)$$

$$= (1-v^2) - \frac{1}{6}z^2v^2(1-v^2) + \frac{1}{360}z^4(8v^2 - 5v^4 - 3v^6) + \dots$$

$$= (1-v^2)\left[1 - \frac{1}{6}z^2v^2\right] + O(z^4) \quad ,$$

$$D_2 = 2\left[\frac{2}{z^2}(\cos zv - \cos z)\right]^{-1}$$

$$= 2\left[\frac{2}{z^2}\left\{1 - \frac{1}{2}z^2v^2 + \frac{1}{24}z^4v^4 - \frac{1}{6!}z^6v^6 + \cdots \right.\right.$$

$$\left.\left. - 1 + \frac{1}{2}z^2 - \frac{1}{24}z^4 + \frac{1}{6!}z^6 + \cdots\cdots \right\}\right]^{-1}$$

$$= 2(1-v^2)^{-1}\left[1 - \frac{z^2}{12}\frac{1-v^4}{1-v^2} + \frac{z^4}{360}\frac{1-v^6}{1-v^2} + \cdots\right]^{-1}$$

$$= 2(1-v^2)^{-1}\left[1 + \frac{z^2}{12}(1+v^2) + O(z^4)\right]$$

So

$$A = \frac{2\sin z}{(1-v^2)}\left(1 - \frac{1}{6}z^2v^2\right)\left[1 + \frac{1}{12}z^2(1+v^2)\right] + O(z^4)$$

$$= \frac{2\sin z}{(1-v^2)}\left[1 + \frac{1}{12}z^2(1-v^2)\right] + O(z^4)$$

and thus

$$CA = \frac{16}{\sin z}A = \frac{2\cdot 16}{(1-v^2)}\left[1 + \frac{1}{12}z^2(1-v^2)\right] + O(z^4) \qquad \text{(E.2)}$$

For the expansion of CB, we need

$$\frac{16}{\sin z} = 16\left(1 + \frac{1}{6}z^2 + \frac{7}{360}z^4 + \cdots\right) \qquad \text{(E.3)}$$

$$2(\cos zv - \cos z) = z^2\left\{(1-v^2) - \frac{1}{12}z^2(1-v^4) + \frac{1}{360}z^4(1-v^6) + \cdots\right\}$$

$$-(1-v^2)z\sin z = z^2\left\{-(1-v^2) + \frac{1}{6}z^2(1-v^2) - \frac{1}{120}z^4(1-v^2) + \cdots\cdots\right\}$$

$$2(\cos zv - \cos z) - (1-v^2)z\sin z = z^2\left\{(1-v^2) - \frac{1}{12}z^2(1-v^4)\right.$$

$$+ \frac{1}{360}z^4(1-v^6) + \cdots\cdots - (1-v^2) + \frac{1}{6}z^2(1-v^2)$$

$$\left. - \frac{1}{120}z^4(1-v^2) + \cdots\cdots\right\}$$

$$= \frac{1}{12}z^4(1-v^2)^2\left\{1 - \frac{1}{30}z^2(v^2+2) + \cdots\right\} \quad ,$$

$$z(\cos z \cdot \sin z - z) = -\frac{2}{3} z^4 \{1 - \frac{1}{5} z^2 + \cdots\}$$

$$\frac{16z}{\sin z} \cdot z(\cos z \cdot \sin z - z) = 16\left(-\frac{2}{3}\right) z^4 \{1 - \frac{1}{30} z^2 + \cdots\}$$

$$\left[2(\cos zv - \cos z) - z(1-v^2)\sin z\right]^{-1} = \frac{12}{z^4(1-v^2)^2} \{1 + \frac{1}{30} z^2(v^2+2) + \cdots\}$$

which, together with (E.2), finally yields

$$K_1 = CA + CB$$

$$= \frac{16}{(1-v^2)^2}\left[-2(3+v^2) + \frac{1}{30} z^2(5v^4 - 18v^2 - 3)\right] + O(z^4) \tag{E.4}$$

Now we set

$$K_2 = \frac{16z}{\sin z} \quad z^2 \frac{2(\cos zv - \cos z) - (1-v^2)\cos z \cdot \sin^2 z}{\left[2(\cos zv - \cos z) - (1-v^2)\, z \sin z\right]^2}$$

$$=: \frac{16z}{\sin z} \quad z^2 A \cdot B$$

with

$$z \sin z = z^2 \{1 - \frac{1}{6} z^2 + \frac{1}{120} z^4 + \cdots\}$$

then yields

$$B \equiv \left[2(\cos zv - \cos z) - (1-v^2)\, z \sin z\right]^{-2}$$

$$= \frac{(12)^2}{z^8(1-v^2)^4} \{1 + \frac{z^2}{15} \frac{2-3v^2+v^6}{(1-v^2)^2} + \cdots\}$$

and from

$$\cos z \cdot \sin^2 z = z^2 \left[1 - \frac{5}{6} z^2 + \frac{91}{360} z^4 + \cdots\right]$$

it follows, together with (E.3), that

$$A \equiv 2(\cos zv - \cos z) - (1-v^2)\cos z \cdot \sin^2 z$$

$$= z^4 \left\{ \frac{1}{12}(9 - 10v^2 + v^4) - \frac{z^2}{360}(90 - 91v^2 + v^6) + \cdots \right\}$$

So the expansion for K_2 reads

$$K_2 = \frac{16 \cdot 12}{z^2(1-v^2)^4}\left(1 + \frac{1}{6}z^2 + \cdots\right)\left\{1 + \frac{1}{15}z^2 \frac{2 - 3v^2 + v^6}{(1-v^2)^2} + \cdots\right\} \cdot$$

$$\cdot \left\{(90 - 10v^2 + v^4) - \frac{z^2}{30}(90 - 91v^2 + v^6) + \cdots\right\}$$

This can be simplified to

$$K_2 = \frac{16 \cdot 12}{z^2} \frac{9 - v^2}{(1-v^2)^2}\left\{\frac{1}{1-v^2} - \frac{1}{30} z^2 + O(z^4)\right\} \quad \text{(E.5)}$$

Last of all, we still need the expansion of ln(b/a), which gives

$$\ln\frac{b}{a} = \ln\left[\frac{4}{1-v^2} \frac{\cos zv - \cos z}{2z \sin z}\right]$$

$$= \ln\left[1 + \frac{1}{12}z^2(1-v^2) + \frac{1}{360}z^4(3-v^2)(1-v^2) + O(z^6)\right]$$

$$= \frac{1}{12}z^2(1-v^2)\left[1 + \frac{z^2}{120}(7+v^2)\right] + O(z^6) \quad ;$$

therefore

$$K_2 \ln\frac{b}{a} = 16\,\frac{9-v^2}{1-v^2}\left[\frac{1}{1-v^2} + \frac{z^2}{120}\,\frac{7+v^2}{1-v^2} - \frac{z^2}{30}\right] + O(z^4)$$

which, together with (E.4), leads to the very simple end result

$$K \equiv K_1 + K_2 \ln\frac{b}{a}$$

$$=: K_{02} + O(z^4)$$

with

$$K_{02} = \frac{48}{1-v^2} + 2z^2 \qquad \text{(E.6)}$$

Now the divergent part of the Laurent series of $K = K(v^2)$ about $v^2 = 1$ must be determined. We begin with

$$K_1 = \frac{16 z^3}{\sin z}\Bigg[\sin z\,\underbrace{\frac{\cos zv - v\cot z \cdot \sin zv}{(1-v^2)(\cos zv - \cos z)}}_{=:\,T_1}$$

$$+ (\cos z \cdot \sin z - z)\,\underbrace{\frac{1}{2(\cos zv - \cos z) - (1-v^2)\,z\sin z}}_{=:\,T_2}\Bigg]$$

and set

$$T_1 \equiv \frac{1}{1-v^2}\,\frac{\cos zv - v\cot z \cdot \sin zv}{\cos zv - \cos z} =: \frac{1}{1-v^2}\,\frac{f_1(v)}{f_2(v)} =: \frac{A_{-1}}{1-v^2}$$

We now need the following values of f_i and f_i'

$$f_1(v) = \cos zv - v\cot z\,\sin zv$$

$$f_1(1) = 0$$

$$f_1'(v) = -z\sin zv - \cot z \sin zv - vz\cot z\cdot\cos zv$$

$$f_1'(1) = -(z\sin z + \cos z + z\cot z\cdot\cos z)$$

$$f_2(v) = \cos zv - \cos z$$

$$f_2(1) = 0$$

$$f_2'(v) = -z\sin zv$$

$$f_2'(1) = -z\sin z$$

Apparently, both f_1 and f_2 have a simple zeroes for $v = 1$; then, with the de L'Hospital rule

$$A_{-1} = \frac{f_1'(1)}{f_2'(1)} = \frac{z\sin z + \cos z + z\cot z\cdot\cos z}{z\sin z}$$

$$= \frac{1}{\sin^2 z} + \frac{\cot z}{z} \quad .$$

This means

$$T_1 = \frac{1}{1-v^2}\left[\frac{1}{\sin^2 z} + \frac{\cot z}{z}\right] + \quad \text{(for } v \to 1 \text{ regular terms).}$$

To evaluate

$$T_2 = \left[2(\cos zv - \cos z) - (1-v^2)\, z\sin z\right]^{-1}$$

we need

$$f(v) \equiv 2(\cos zv - \cos z) - (1-v^2)\, z\sin z$$

$$f(1) = 0$$

$$f'(v) = -2z\sin zv + 2zv\sin z$$

$$f'(1) = 0$$

$$f''(v) = -2z^2\cos zv + 2z\sin z$$

$$f''(1) = -2z^2\cos z + 2z\sin z$$

$$f'''(v) = 2z^3\sin zv$$

$$f'''(1) = 2z^3\sin z$$

Because of $f''(1) \neq 0$, T_2 has a second order pole for $v \to 1$, i.e., an expansion exists in the form

$$T_2 = \frac{A'_{-2}}{(1-v^2)^2} + \frac{A'_{-1}}{1-v^2} + A'_0 + \cdots ,$$

whereby we shall in the following only concern ourselves with the coefficients A'_{-2} and A'_{-1}. The de L'Hospital rule now gives

$$A'_{-2} = \lim_{v\to 1} (1-v^2)^2 T_2$$

$$= \lim_{v\to 1} \frac{(1-v^2)^2}{f(v)}$$

$$= \frac{g''(1)}{f''(1)}$$

with $g(v) := (1-v^2)^2$; taking the derivative, we get

$$g(v) \equiv (1-v^2)^2$$

$$g(1) = 0$$

$$g'(v) = -4v(1-v^2)$$

$$g'(1) = 0$$

$$g''(v) = -4(1-v^2)+8v^2$$
$$g''(1) = 8$$
$$g'''(v) = 24v$$
$$g'''(1) = 24$$

Then

$$A'_{-2} = \frac{g''(1)}{f''(1)} = \frac{4}{z(\sin z - z\cos z)}$$

We calculate the next coefficient from

$$A'_{-1} = \lim_{v\to 1} (1-v^2)\left[T_2 - \frac{A'_{-2}}{(1-v^2)^2}\right]$$
$$= \lim_{v\to 1} \left[T_2(1-v^2) - \frac{A'_{-2}}{1-v^2}\right]$$
$$= \lim_{v\to 1} \left[\frac{1-v^2}{2(\cos zv - \cos z)-(1-v^2)z\sin z} - \frac{A'_{-2}}{1-v^2}\right]$$
$$= \lim_{v\to 1} \frac{(1-v^2)^2 - A'_{-2}\left[2(\cos zv-\cos z)-(1-v^2)z\sin z\right]}{(1-v^2)\left[2(\cos zv-\cos z)-(1-v^2)z\sin z\right]}$$
$$= \lim_{v\to 1} \frac{g(v) - A'_{-2}f(v)}{(1-v^2)f(v)}$$

Numerator and denominator have a three-fold zero so that

$$A'_{-1} = \frac{g'''(1) - A'_{-2}f'''(1)}{\left[(1-v^2)f(v)\right]'''(v=1)}$$

The denominator of this is

$$\left[(1-v^2)f(v)\right]'''(v=1) = \Big[(1-v^2)'''f(v) + 3(1-v^2)''f'(v) +$$

$$+ 3(1-v^2)'f''(v) + (1-v^2)f'''(v)\Big](v=1)$$

$$= -6f''(1)$$

$$= -12\,z(\sin z - z\cos z)$$

Then

$$A'_{-1} = \frac{24 - \dfrac{4}{z(\sin z - z\cos z)}\,2z^3\sin z}{-12\,z(\sin z - z\cos z)}$$

$$= \frac{2}{3}\;\frac{3(z\cos z - \sin z) + z^2\sin z}{z(z\cos z - \sin z)^2}$$

We now write

$$K_1 = 16\,z^2\left[T_1 + \frac{\cos z\cdot\sin z - z}{\sin z}\,T_2\right]$$

$$=: \frac{k_{-2}}{(1-v^2)^2} + \frac{k_{-1}}{(1-v^2)} + k_0 + \cdots$$

and get

$$k_{-2} = \frac{64\,z}{\sin z}\;\frac{\cos z\cdot\sin z - z}{\sin z - z\cos z}$$

$$k_{-1} = 16\,z^2\left[\frac{1}{\sin^2 z} + \frac{\cot z}{z} - 2\,\frac{\cos z\cdot\sin z - z}{z\sin^2 z\,(1 - z\cot z)}\right.$$

$$\left. + \frac{2}{3}\,\frac{z(\cos z\cdot\sin z - z)}{\sin^2 z\,(1 - z\cot z)}\right]. \qquad \text{(E.7)}$$

Now we examine the expression

$$\frac{b}{a} = \frac{2(\cos zv - \cos z)}{(1-v^2)\,z\sin z}$$

which, for $v = 1$, exhibits in the numerator and denominator a first order zero; this means that it can be expanded in a power series of the general form

$$\frac{b}{a} = l_0 + l_1(1-v^2) + l_2(1-v^2)^2 + \ldots .$$

As will be shown, we must calculate the coefficients l_o, l_1 and l_2 of this series. Again, with the de L'Hospital rule,

$$l_0 = \lim_{v\to 1} \frac{2(\cos zv - \cos z)}{(1-v^2)\, z \sin z}$$

$$= \frac{[2(\cos zv - \cos z)]'(v=1)}{[(1-v^2)\, z \sin z]'(v=1)} \qquad (E.8)$$

$$= \frac{-2z \sin z}{-2z \sin z} = 1 ,$$

as well as

$$l_1 = \lim_{v\to 1} \left[\frac{b}{a} - l_0\right] \frac{1}{1-v^2}$$

$$= \lim_{v\to 1} \frac{2(\cos zv - \cos z) - (1-v^2)\, z \sin z}{(1-v^2)^2\, z \sin z}$$

$$= \frac{f''(1)}{g''(1)\, z \sin z} \qquad (E.9)$$

$$= \frac{2z(\sin z - z\cos z)}{8z \sin z} = \frac{1}{4}(1 - z \cot z)$$

and

$$l_2 = \lim_{v\to 1} \left[\frac{b}{a} - l_0 - l_1(1-v^2)\right] \frac{1}{(1-v^2)^2}$$

$$= \lim_{v\to 1} \left[\frac{2(\cos zv - \cos z)}{(1-v^2)^3\, z \sin z} - \frac{1}{(1-v^2)^2} - \frac{l_1}{(1-v^2)}\right]$$

$$= \lim_{v\to 1} \frac{2(\cos zv - \cos z) - (1-v^2)\, z \sin z - l_1 (1-v^2)^2\, z \sin z}{(1-v^2)^3\, z \sin z}$$

$$= \lim_{v \to 1} \frac{f(v) - \ell_1\, g(v)\, z \sin z}{h(v) \quad z \sin z}$$

$$= \frac{f'''(1) - \ell_1\, g'''(1)\, z \sin z}{h'''(1)\, z \sin z}$$

Here,

$$h(v) := (1-v^2)^3$$

$$h(1) = 0$$

$$h'(v) = -6v(1-v^2)^2$$

$$h'(1) = 0$$

$$h''(v) = -6(1-v^2)^2 + 24v^2(1-v^2)$$

$$h''(1) = 0$$

$$h'''(v) = 3 \cdot 24\, v(1-v^2) - 48\, v^3$$

$$h'''(1) = -48$$

$$h''''(v) = 3 \cdot 24(1-v^2) - 6 \cdot 48 v^2$$

$$h''''(1) = -6 \cdot 48$$

It follows that

$$\ell_2 = \frac{2z^3 \sin z - \frac{1}{4}(1 - z\cot z) \cdot 24\, z \sin z}{(-48)\, z \sin z}$$

$$= -\frac{1}{24}\left[z^2 - 3(1 - z\cot z)\right] \qquad \text{(E.10)}$$

With $\varepsilon := 1 - v^2$, the power series for ln(b/a) reads

$$\ln\frac{b}{a} = \ln(1 + \ell_1\varepsilon + \ell_2\varepsilon^2 + \dots)$$

$$= \ell_1\varepsilon + \ell_2\varepsilon - \frac{1}{2}(\ell_1\varepsilon)^2 + O(\varepsilon^3)$$

$$= \ell_1 \varepsilon + (\ell_2 - \tfrac{1}{2}\ell_1)\varepsilon^2 + O(\varepsilon^3)$$

$$=: L_1 \varepsilon + L_2 \varepsilon^2 + O(\varepsilon^3)$$

Here,

$$L_1 \equiv \ell_1 = \frac{1}{4}(1 - z\cot z) \tag{E.11}$$

and

$$L_2 \equiv \ell_2 - \tfrac{1}{2}\ell_1$$

$$= -\frac{1}{96}\left\{4z^2 - 12(1 - z\cot z) + 3(1 - z\cot z)^2\right\} . \tag{E.12}$$

Now we set

$$K_2 = \frac{16z^3}{\sin z} \; \frac{2(\cos zv - \cos z) - (1 - v^2)\cos z \cdot \sin^2 z}{[2(\cos zv - \cos z) - (1 - v^2)\, z \sin z]^2}$$

$$=: \frac{16z^3}{\sin z} \; T_3$$

$$=: \frac{16z^3}{\sin z} \; \frac{\phi(v)}{f^2(v)}$$

and get

$$\begin{aligned}
\phi(v) &:= 2(\cos zv - \cos z) - (1 - v^2)\cos z \cdot \sin^2 z \\
\phi(1) &= 0 \\
\phi'(v) &= -2z\sin zv + 2v\cos z \cdot \sin^2 z \\
\phi'(1) &= 2\sin z\,(\cos z \cdot \sin z - z) \\
\phi''(v) &= -2z^2\cos zv + 2\cos z \cdot \sin^2 z \\
\phi''(1) &= 2\cos z\,(\sin^2 z - z^2) \\
\phi'''(v) &= 2z^3 \sin zv \\
\phi'''(1) &= 2z^3 \sin z \\
\phi''''(v) &= 2z^4 \cos zv \\
\phi''''(1) &= 2z^4 \cos z
\end{aligned}$$

Apparently the numerator of T_3 has a single zero, while the denominator has a four-fold zero at v = 1. Therefore the Laurent series of T_3 takes the form

$$T_3 = \frac{C_{-3}}{\varepsilon^3} + \frac{C_{-2}}{\varepsilon^2} + \frac{C_{-1}}{\varepsilon} + C_0 + \cdots$$

Thus

$$T_3 \ln\frac{b}{a} = \left(\frac{C_{-3}}{\varepsilon^3} + \frac{C_{-2}}{\varepsilon^2} + \frac{C_{-1}}{\varepsilon} + \cdots\right)\varepsilon\{L_1 + L_2\varepsilon + \cdots\}$$

$$= \frac{L_1 C_{-3}}{\varepsilon^2} + \frac{L_2 C_{-3} + L_1 C_{-2}}{\varepsilon} + O(\varepsilon^0)$$

and

$$K_2 \ln\frac{b}{a} = \frac{16z^3}{\sin z}\left[\frac{L_1 C_{-3}}{\varepsilon^2} + \frac{L_2 C_{-3} + L_1 C_{-2}}{\varepsilon} + \cdots\right]$$

$$=: \frac{\varkappa_{-2}}{(1-v^2)^2} + \frac{\varkappa_{-1}}{1-v^2} + \cdots \qquad \text{(E.13)}$$

with

$$\varkappa_{-2} \equiv \frac{16z^3}{\sin z} L_1 C_{-3}$$

$$\varkappa_{-1} \equiv \frac{16z^3}{\sin z}\left[L_2 C_{-3} + L_1 C_{-2}\right] \qquad \text{(E.14)}$$

This means that we must explicitly calculate the coefficients C_{-2} and C_{-3}. By means of our usual method,

$$C_{-3} = \lim_{v\to 1} (1-v^2)^3 T_3$$

$$= \lim_{v\to 1} \frac{h(v)\,\phi(v)}{f^2(v)}$$

$$= \frac{(h\phi)''''(v=1)}{(f^2)''''(v=1)}$$

The necessary derivatives are

$$(h\phi)''''(v=1) = \left[h''''\phi + 4h'''\phi' + 6h''\phi'' + 4h'\phi''' + h\phi''''\right](v=1)$$

$$= 4\,h'''(1)\,\phi'(1)$$

$$= -\,8\cdot 48\,\sin z\,(\cos z\cdot\sin z - z)$$

$$(f^2)''''(v=1) = \left[f''''f + 4f'''f' + 6f''f'' + 4f'f''' + ff''''\right](v=1)$$

$$= 6\,(f''(1))^2$$

$$= 24\,z^2\,(\sin z - z\cos z)^2$$

so that

$$C_{-3} = \frac{-8\cdot 48\,\sin z\,(\cos z\cdot\sin z - z)}{24\,z^2\,(\sin z - z\cos z)^2}$$

$$= -\,\frac{16}{z^2}\,\frac{\cos z\cdot\sin z - z}{\sin z\,(1 - z\cot z)^2}\ .$$

Accordingly

$$C_{-2} = \lim_{v\to 1}\left[T_3 - \frac{C_{-3}}{(1-v^2)^3}\right](1-v^2)^2$$

$$= \lim_{v\to 1}\left[T_3(1-v^2)^2 - \frac{C_{-3}}{1-v^2}\right]$$

$$= \lim_{v\to 1}\left[\frac{(1-v^2)^2\phi(v)}{f^2(v)} - \frac{C_{-3}}{1-v^2}\right]$$

$$= \lim_{v\to 1}\frac{(1-v^2)^3\phi(v) - C_{-3}f^2(v)}{(1-v^2)f^2(v)}$$

$$= \lim_{v\to 1}\frac{h(v)\,\phi(v) - C_{-3}f^2(v)}{(1-v^2)\,f^2(v)}$$

$$= \frac{(h\phi)'''''(v=1) - C_{-3}\,(f^2)'''''(v=1)}{\left[(1-v^2)f^2\right]'''''(v=1)}$$

The derivatives are

$$(h\phi)'''''(1) = [h'''''\phi + 5h''''\phi' + 10h'''\phi'' + 10h''\phi''' + 5h'\phi'''' + h\phi'''''](1)$$
$$= 5(-6\cdot 48)\phi'(1) - 480\,\phi''(1)$$
$$= -480\,(3\,\phi'(1) + \phi''(1))$$

$$(f^2)'''''(1) = [f'''''f + 5f''''f' + 10f'''f'' + 10f''f''' + 5f'f'''' + ff'''''](1)$$
$$= 20 f''(1) f'''(1)$$
$$= 80\, z^4 \sin z\,(\sin z - z\cos z)$$

With

$$(f^2)'''(1) = [f'''f + 3f''f + 3f'f'' + ff'''](1) = 0$$

and

$$(1-v^2)'(1) = -2$$
$$(1-v^2)''(1) = -2$$
$$(1-v^2)^{(n)}(1) = 0 \quad , \forall\, n \geq 3$$

leading to

$$[(1-v^2)f^2]'''''(1) = [(1-v^2)'''''f^2 + 5(1-v^2)''''f^{2'} + 10\,(1-v^2)'''f^{2''}$$
$$+ 10\,(1-v^2)''f^{2'''} + 5(1-v^2)'f^{2''''} + (1-v^2)f^{2'''''}](1)$$
$$= 5\,(1-v^2)'(1)\,f^{2''''}(1)$$
$$= -240\, z^2(\sin z - z\cos z)^2$$

If we put all of this into C_{-2}, we get

$$C_{-2} = [-240\, z^2(\sin z - z\cos z)^2]^{-1}\,\{-480\,[6\sin z(\cos z\cdot\sin z - z) +$$
$$+ 2\cos z\,(\sin^2 z - z^2)] -$$

$$- \frac{(-16)\, 80 \sin z \,(\cos z \cdot \sin z - z) \cdot z^4 \sin z\,(\sin z - z\cos z)}{z^2\,(\sin z - z\cos z)^2}$$

$$= 4\, \frac{3 \sin z\,(\cos z \cdot \sin z - z) + \cos z\,(\sin^2 z - z^2)}{z^2 \sin^2 z\,(1 - z\cot z)^2}$$

$$- \frac{16}{3}\, \frac{\cos z \cdot \sin z - z}{\sin z\,(1 - z\cot z)^3}$$

We can now calculate the κ_i with (E.14); it follows for κ_{-2}

$$\varkappa_{-2} = \frac{16 z^3}{\sin z}\, L_1 C_{-3}$$

$$= \frac{16 z^3}{\sin z}\, \frac{1}{4}(1 - z\cot z)\frac{(-16)}{z^2}\, \frac{\sin z\,(\cos z \cdot \sin z - z)}{\sin^2 z\,(1 - z\cot z)^2}$$

$$= - \frac{64 z}{\sin z}\, \frac{\cos z \cdot \sin z - z}{\sin z - z\cos z}$$

which, together with (E.7) leads to

$$k_{-2} + \varkappa_{-2} = 0\,,$$

i.e., K(z,v) has only a simple pole at v = 1! (This fact is decisive for the renormalizability of $L^{(2)}$).

For κ_{-1} we need

$$L_2 C_{-3} = - \frac{1}{96}\left(-\frac{16}{z^2}\right)\left\{4z^2 - 12(1 - z\cot z) + 3(1 - z\cot z)^2\right\}\frac{(\cos z \cdot \sin z - z)}{\sin z\,(1 - z\cot z)^2}$$

$$= \frac{2}{3}\, \frac{\cos z \cdot \sin z - z}{\sin z\,(1 - z\cot z)^2} - \frac{2}{z^2}\, \frac{\cos z \cdot \sin z - z}{\sin z\,(1 - z\cot z)} + \frac{\cos z \cdot \sin z - z}{2 \sin z}$$

and

$$L_1 C_{-2} = \frac{1}{4}(1 - z\cot z)\, C_{-2}$$

$$= \frac{3\sin z\,(\cos z\cdot\sin z - z) + \cos z\,(\sin^2 z - z^2)}{z^2\,\sin^2 z\,(1 - z\cot z)} - \frac{4}{3}\,\frac{\cos z\cdot\sin z - z}{\sin z\,(1 - z\cot z)^2}$$

the sum is

$$L_2C_{-3} + L_1C_{-2} = -\frac{2}{3}\,\frac{\cos z\cdot\sin z - z}{\sin z\,(1 - z\cot z)^2} + \frac{\cos z\cdot\sin z - z}{z^2\sin z\,(1 - z\cot z)}$$

$$+ \frac{\cos z\cdot\sin z - z}{2z^2\,\sin z} + \frac{\cos z\,(\sin^2 z - z^2)}{z^2\sin^2 z\,(1 - z\cot z)}$$

so that it follows that

$$\mathcal{K}_{-1} = \frac{16\,z^3}{\sin z}\left(L_2C_{-3} + L_1C_{-2}\right)$$

$$= 16z^2\Big[-\frac{2}{3}\,\frac{z(\cos z\cdot\sin z - z)}{\sin^2 z\,(1 - z\cot z)^2} + \frac{\cos z\cdot\sin z - z}{z\cdot\sin^2 z\,(1 - z\cot z)}$$

$$+ \frac{\cos z\cdot\sin z - z}{2z\,\sin^2 z} + \frac{\cos z\,(\sin^2 z - z^2)}{z\,\sin^3 z\,(1 - z\cot z)}\Big]$$

$$= 16z^2\Big[-\frac{2}{3}\,\frac{z(\cos z\cdot\sin z - z)}{\sin^2 z\,(1 - z\cot z)^2} + \frac{(\cos z\cdot\sin z - z)[3 - z\cot z]}{2z\,\sin^2 z\,(1 - z\cot z)}$$

$$+ \frac{\cos z\,(\sin^2 z - z^2)}{z\,\sin^3 z\,(1 - z\cot z)}\Big].$$

From this, we get with (E.7)

$$\mathcal{K}_{-1} + k_{-1} = 16z^2\Big[\frac{1}{\sin^2 z} + \frac{\cot z}{z} \underbrace{- 2\,\frac{\cos z\cdot\sin z - z}{z\,\sin^2 z\,(1 - z\cot z)}}_{=:P} + \frac{2}{3}\,\frac{z(\cos z\cdot\sin z - z)}{\sin^2 z\,(1 - z\cot z)^2}$$

$$-\frac{2}{3}\,\frac{z(\cos z\cdot\sin z - z)}{\sin^2 z\,(1 - z\cot z)^2} + \underbrace{\frac{(\cos z\cdot\sin z - z)[3 - z\cot z]}{2z\,\sin^2 z\,(1 - z\cot z)}}_{=:Q} + \underbrace{\frac{\cos z\cdot(\sin^2 z - z^2)}{z\,\sin^3 z\,(1 - z\cot z)}}_{=:R}\Big]$$

This expression can be considerably simplified:

$$P + Q = -\frac{1}{2}\,\frac{\cos z\cdot\sin z - z}{z\,\sin^2 z}\cdot\frac{1 + z\cot z}{1 - z\cot z},$$

$$P+Q+R=\{2z\sin^2 z\,(1-z\cot z)\}^{-1}[2\cot z\,\sin^2 z-2z^2\cot z$$
$$-(\cos z\cdot\sin z-z)(1+z\cot z)]$$
$$=\{\cdots\}^{-1}[2\cot z\cdot\sin^2 z-z^2\cot z-\cos z\cdot\sin z-z\cot z\cdot\cos z\cdot\sin z+z]$$
$$=\{\cdots\}^{-1}[\sin z\cdot\cos z\,(1-z\cot z)+z(1-z\cot z)]$$
$$=\{\cdots\}^{-1}(z+\sin z\cos z)(1-z\cot z)$$
$$=\frac{z+\sin z\cos z}{2z\sin^2 z}$$
$$=\frac{1}{2}\left\{\frac{1}{\sin^2 z}+\frac{\cot z}{z}\right\}$$

All together, then,

$$\mathcal{K}_{-1}+k_{-1}=16z^2\left[\frac{3}{2}\frac{1}{\sin^2 z}+\frac{3}{2}\frac{\cot z}{z}\right]$$
$$=24\left[\frac{z^2}{\sin^2 z}+z\cot z\right]$$

This is the desired result which we can also write in the form

$$K(v)=\frac{24}{1-v^2}\left[\frac{z^2}{\sin^2 z}+z\cot z\right]\quad+\text{ (regular terms).}$$

Appendix F: Contact Term Determination in Source Theory

In the following, the calculation of the two-loop effective Lagrangian in the 7th section is interpreted in the framework of Schwinger's source theory [29,30,47]. In this phenomenological theory, there is no difference made between bare and renormalized quantities; all masses, charges, field strengths etc. which appear are the physical, i.e., observable parameters. Applied to QED, this means, for example, that the fermions should always propagate with their observable mass m, thereby guaranteeing that one substitutes g(p) + ct for the propagator g(p) and defines the contact terms ct at the end of all calculations in such a manner that the pole of g is given by m. On the level of the mass operator, this 'on-shell'-condition is fulfilled by the substitution

$$\Sigma(p) \longrightarrow \Sigma(p) + ct$$

The ct are to be determined thereby from

$$\Sigma(\not p = -m) = 0$$

$$\frac{\partial \Sigma}{\partial \not p}(\not p = -m) = 0$$

Analogously, for the polarization tensor, we set

$$\Pi_{\mu\nu}(k) \longrightarrow \Pi_{\mu\nu}(k) + ct$$

and require that $\Pi_{\mu\nu}(k^2 = 0) = 0$. Our considerations in the third and fourth section show that this is equivalent to the procedure in conventional operator field theory.

In Source Theory, we now replace (7.10) by

$$\mathcal{L}^{(2)} = \frac{e^2}{2}\int\frac{d^4k}{(2\pi)^4}\int\frac{d^4p}{(2\pi)^4}\, tr\,[\gamma_\mu(g(p)+ct_1^G)\gamma^\mu \cdot (g(p-k)+ct_2^G)]\,D_+(k) + ct' \tag{F.1}$$

and define ct_i^G in such a way that the fermions always propagate with their physical mass; the ct's result from the requirements

$$\text{(i)}\quad \mathcal{L}^{(2)}(B=0) = 0$$
$$\text{(ii)}\quad \mathcal{L}^{(2)}(B) < \infty\ ,\quad \forall\, B \geq 0 \tag{F.2}$$

By means of the substitution $p =: p' + \frac{k}{2}$, we can write (F.1) somewhat more symmetrically:

$$\begin{aligned}\mathcal{L}^{(2)} &= \frac{e^2}{2}\int\frac{d^4k}{(2\pi)^4}\int\frac{d^4p'}{(2\pi)^4}\, tr[\gamma_\mu(g(p'+\tfrac{k}{2})+ct_1^G)\gamma^\mu(g(p'-\tfrac{k}{2})+ct_2^G)]D_+(k)+ct' \\ &= \underbrace{\frac{e^2}{2}\int\frac{d^4k}{(2\pi)^4}\int\frac{d^4p'}{(2\pi)^4}\, tr[\gamma_\mu\, g(p'+\tfrac{k}{2})\gamma^\mu g(p'-\tfrac{k}{2})]\,D_+(k)+ct'}_{=:\ \tilde{\mathcal{L}}^{(2)}} \\ &\quad + \frac{e^2}{2}\int\frac{d^4k}{(2\pi)^4}\int\frac{d^4p'}{(2\pi)^4}\, tr[\gamma_\mu\, g(p'+\tfrac{k}{2})\gamma^\mu\, ct_2^G]\,D_+(k) \\ &\quad + \frac{e^2}{2}\int\frac{d^4k}{(2\pi)^4}\int\frac{d^4p'}{(2\pi)^4}\, tr[\gamma_\mu\, g(p'-\tfrac{k}{2})\gamma^\mu\, ct_1^G]\,D_+(k) \\ &=: \tilde{\mathcal{L}}^{(2)} + ct_2^L + ct_1^L\end{aligned} \tag{F.3}$$

At the end, a divergent constant was incorporated into ct'. The symmetry of $L^{(2)}$ in both electron propagators leads to $ct_1^G = ct_2^G$, which because of (F.3) implies $ct_1^L = ct_2^L$. Thus it follows that

$$\mathcal{L}^{(2)} = \tilde{\mathcal{L}}^{(2)} + 2\,ct_2^L\ ,$$

i.e., we only have to calculate the additional term of <u>one</u> electron to $L^{(2)}$ and can multiply it with 2 to get $ct_1^L + ct_2^L$. We now use the fact that we can also write $L^{(2)}$ as a folding of an electron propagator with the mass operator calculated to the order α

$$\mathcal{L}^{(2)} = -\frac{i}{2}\int\frac{d^4p}{(2\pi)^4}\,\mathrm{tr}\left\{ g(p)\left[ie^2\int\frac{d^4k}{(2\pi)^4}\gamma^\mu g(p-k)\gamma_\mu D_+(k)\right]\right\}$$

$$= -\frac{i}{2}\int\frac{d^4p}{(2\pi)^4}\,\mathrm{tr}\left\{ g(p)\Sigma(p)\right\}$$

We again introduce contact terms

$$\mathcal{L}^{(2)} \overset{?}{=} -\frac{i}{2}\int\frac{d^4p}{(2\pi)^4}\,\mathrm{tr}\left\{ g(p)\left(\Sigma(p)+ct_\Sigma\right)\right\} + ct' .$$

The ct_Σ are only responsible for $\Sigma(\not p = -m) = 0$, since this conforms to a propagation with the physical mass, i.e., with the notation from section 3 or 7, ct_Σ is equal to $-\delta m$. The above formula is not yet correct, since ct_Σ only takes the mass displacement of one electron of the loop (namely, of the electron in Σ) into account; to be correct, it would have to read

$$\mathcal{L}^{(2)} = -\frac{i}{2}\int\frac{d^4p}{(2\pi)^4}\,\mathrm{tr}\left\{ g(p)\Sigma(p)\right\}$$
$$-\frac{i}{2}\left(2\,ct_\Sigma\right)\mathrm{tr}\int\frac{d^4p}{(2\pi)^4}\, g(p) + ct'$$
$$= \frac{i}{2}\int\frac{d^4k}{(2\pi)^4} D_+(k)\,\Pi_\mu^\mu(k) + ct' - ct_\Sigma\, i\,\mathrm{tr}\, G_+(x,x|A)$$
$$\underset{(5.13)}{=} \tilde{\mathcal{L}}^{(2)} + ct_\Sigma\,\frac{\partial\mathcal{L}^{(1)}}{\partial m} . \qquad \text{(F.4)}$$

The calculation of $L^{(2)}$ is now accomplished exactly like in the 7th cahpter; but now, e and B are not renormalized. Instead, in order to fulfill (F.2), ct' is appropriately chosen.

(m, e, B are now the observable quantities in the entire calculation!) This leads precisely to the subraction of K_{02}, after the singular part of the Laurent series has been removed from the integral. From our earlier results, we get for (F.4): (note that $ct_{\Sigma} = -\delta m$)

$$
\begin{aligned}
\mathcal{L}^{(2)} = \Big[& \frac{\alpha}{(4\pi)^3} \int_0^\infty \frac{ds}{s^3} \int_0^1 dv\, e^{-im^2 s} \left[K(z,v) - K_{02}(z,v) - \frac{f(z)}{1-v^2} \right] \\
& - \frac{5}{6} \frac{3\alpha m}{4\pi} \frac{(-im)}{4\pi^2} \int_0^\infty \frac{ds}{s^2} e^{-im^2 s} \left[(eBs)\cot(eBs) + \frac{1}{3}(eBs)^2 - 1 \right] \\
& + \frac{3\alpha}{16\pi^3} \int_0^\infty \frac{ds}{s^3} e^{-im^2 s} \ln(i\gamma m^2 s) \left[\frac{(eBs)^2}{\sin^2(eBs)} + (eBs)\cot(eBs) - 2 \right] \\
& \cancel{- ct_{\Sigma} \frac{\partial \mathcal{L}^{(1)}}{\partial m}} \Big] + \cancel{ct_{\Sigma} \frac{\partial \mathcal{L}^{(1)}}{\partial m}}
\end{aligned}
$$

This, however, is identical to the result achieved by conventional renormalization (7.37), i.e., in this example, too, Source Theory and Operator Field Theory are equivalent!

Appendix G: One-Loop Effective Lagrangian as Perturbation Series

In this appendix, the expression used in section 5

$$iW^{(1)}[A] = -\,\mathrm{Tr}\,\ln\,(1 - e\not{A}G_+)^{-1}$$

for the one-loop effective action will be derived without the use of functional methods [41], i.e., by calculating all relevant graphs in this approximation with the well-known Feynman rules and then summing them up properly. For that purpose, let us first consider the exact effective action $\Gamma[A]$, which represents the generating functional of the one-particle-irreducible (1PI), amputated N-point functions $\Gamma^{(N)}(x_1,\ldots,x_N)$; i.e., if we take the derivatives of this functional N times with respect to A and then set $A = 0$, then we get the corresponding Green's function. Conversely, $\Gamma[A]$ must then be able to be written as a sum over the $\Gamma^{(N)}$

$$\Gamma[A] = \sum_N \frac{1}{N!} \int d^4x_1 \ldots d^4x_N \; \Gamma^{(N)}_{\mu_1\ldots\mu_N}(x_1\ldots x_N)\, A^{\mu_1}(x_1)\ldots A^{\mu_N}(x_N)$$

This is still an exact equation, in which the $\Gamma^{(N)}$ contain graphs of an arbitrary high order in e^2. If we wish to evaluate the above sum only in one-loop approximation, then the only term contributing to $\Gamma^{(N)}$ is represented by the graph

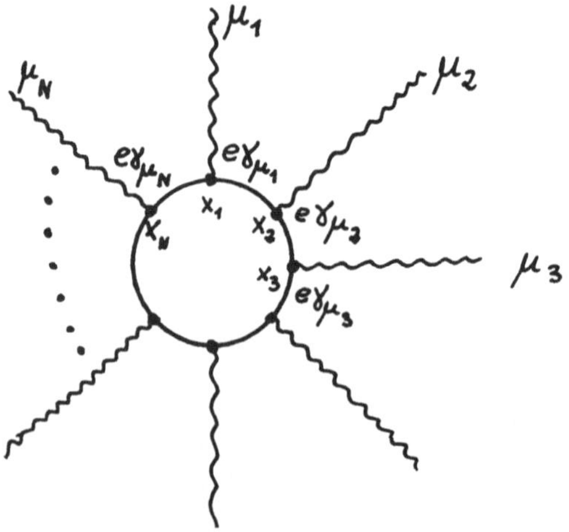

Expressed by formulae, this means

$$\Gamma^{(N)}_{\mu_1\cdots\mu_N}(x_1,\ldots,x_N) = (-1)(N-1)!\ tr\left[(e\gamma_{\mu_1})G_+(x_1,x_2)(e\gamma_{\mu_2})G_+(x_2,x_3)\cdot\right.$$

$$\left.\cdot(e\gamma_{\mu_3})G_+(x_3,x_4)\ldots\ldots(e\gamma_{\mu_N})G_+(x_N,x_1)\right].$$

The individual factors have the following origin:

(1) The factor (-1) appears since a closed Fermion loop is present.

(2) The factor (N-1)! takes into account that, after fixing one photon line, a total of (N-1)! topologically inequivalent graphs can be created by permutation of the remaining photon lines.

(3) One factor $(e\gamma_\mu)$ is assigned to each vertex,

(4) A free propagator G_+ is assigned to each fermion line.

(5) The fermion loop requires a Dirac-trace.

Furthermore, it should be noted that $\Gamma^{(N)}$ does not, by definition, contain the external propagators. For $\Gamma[A]$ in one-loop approximation, i.e., $W^{(1)}[A]$, we thus get

$$\Gamma[A] \approx W^{(1)}[A] = \frac{(-1)}{i}\sum_{N=1}^{\infty}\frac{(N-1)!}{N!}\int d^4x_1\ldots d^4x_N\ tr\left[(e\gamma_{\mu_1})G_+(x_1,x_2)\ldots\right.$$

$$\left.\ldots(e\gamma_{\mu_N})G_+(x_N,x_1)\right]A^{\mu_1}(x_1)\,A^{\mu_2}(x_2)\ldots\,A^{\mu_N}(x_N)$$

$$= \frac{(-1)}{i}\sum_{N=1}^{\infty}\frac{1}{N}\int d^4x_1\ldots d^4x_N\ tr\left[e\not{A}(x_1)G_+(x_1,x_2)\cdot\right.$$

$$\left.\cdot\, e\not{A}(x_2)G_+(x_2,x_3)\ldots\ldots e\not{A}(x_N)G_+(x_N,x_1)\right].$$

(The factor $\frac{1}{i}$ in fromt of the sum follows from the definition $Z = \exp(iW)$.) Furthermore, it follows in matrix notation

$$W^{(1)}[A] = -\frac{1}{i}\, tr \int d^4x_1 \sum_{N=1}^{\infty} \frac{1}{N} \int d^4x_2 \ldots d^4x_N \langle x_1 | e\not{A} G_+ | x_2 \rangle \cdot$$

$$\cdot \langle x_2 | e\not{A} G_+ | x_3 \rangle \cdots \langle x_N | e\not{A} G_+ | x_1 \rangle$$

$$= -\frac{1}{i}\, tr \int d^4x_1 \sum_{N=1}^{\infty} \frac{1}{N} \langle x_1 | (e\not{A} G_+)^N | x_1 \rangle$$

$$= -\frac{1}{i}\, tr \int d^4x_1 \langle x_1 | \sum_{N=1}^{\infty} \frac{1}{N!} (e\not{A} G_+)^N | x_1 \rangle$$

$$= +\frac{1}{i}\, tr \int d^4x_1 \langle x_1 | \ln(1 - e\not{A} G_+) | x_1 \rangle$$

$$= \frac{1}{i}\, \mathrm{Tr} \ln(1 - e\not{A} G_+)$$

Here, we have used the series

$$\ln(1-x) = -\sum_{n=1}^{\infty} \frac{1}{n} x^n \,, \quad x \in (-1,1)$$

We can also write our result as

$$iW^{(1)}[A] = -\mathrm{Tr} \ln(1 - e\not{A} G_+)^{-1}$$

i.e., we obtain the same expression as the one Fried [41] derived with functional methods, qed.

Appendix H: Summary of the Most Important Formulae

Here we summarize the most important formulae of the individual sections in synoptical form; explanations of these equations are to be found in the respective chapters.

2nd Section

(2.1) Definition of the fermion propagator

$$G_+ = \frac{1}{\gamma\pi + m - i\varepsilon} \quad , \ \varepsilon > 0 \ , \ \pi = p - eA$$

(2.6) $$(\gamma\pi)^2 = -\pi^2 + \frac{e}{2}\sigma_{\mu\nu}F^{\mu\nu}$$

especially for $\vec{B} = B\hat{z}$:

$$(\gamma\pi)^2 = -\pi^2 + eB\sigma^3$$

(2.12) Differential equation for G_+ in space representation

$$\left[m + \gamma^\mu\left(\frac{1}{i}\partial'_\mu - eA_\mu(x')\right)\right] G_+(x',x'') = \delta(x' - x'')$$

(2.47) Electron propagator in an external constant magnetic field

$$G_+(x',x'') = \phi(x',x'') \int \frac{d^4k}{(2\pi)^4} e^{ik(x'-x'')} g(k)$$

$$\phi(x',x'') = \exp\left[ie\int_{x''}^{x'} dx_\mu A^\mu(x)\right] \quad \text{(straight path of integration)}$$

$$g(k) = i\int_0^\infty ds \exp\left\{-is\left[m^2 + k_\parallel^2 + (\tan z/z)\, k_\perp^2 - i\varepsilon\right]\right\} \cdot \frac{e^{i\sigma^3 z}}{\cos z}\left(m - \gamma k_\parallel - \frac{e^{-i\sigma^3 z}}{\cos z}\gamma k_\perp\right)$$

3rd Section

(3.5) Space representation of the mass operator (Feynman Gauge)

$$\Sigma(x',x'') = ie^2\gamma^\mu G_+(x',x'') D_+(x'-x'')\gamma_\mu + O(e^4)$$

(3.6) free photon propagator

$$D_+(x) = \int \frac{d^4k}{(2\pi)^4} e^{ikx} \frac{1}{k^2 - i\varepsilon}$$

(3.8) transition to momentum representation

$$\Sigma(x',x'') = \phi(x',x'') \int \frac{d^4p}{(2\pi)^4} e^{ip(x'-x'')} \Sigma(p)$$

$$(3.9)\quad \Sigma(p) = ie^2\gamma^\mu \int \frac{d^4k}{(2\pi)^4} \frac{1}{k^2 - i\varepsilon} g(p-k)\gamma_\mu + O(e^4)$$

(3.44) Tsai's representation of the mass operator in a constant external magnetic field

$$\Sigma(\Pi) = \frac{\alpha m}{2\pi} \int_0^\infty \frac{ds}{s} \int_0^1 du\, e^{-isu^2m^2} \Big\{ \Delta^{-\frac{1}{2}} e^{-is\phi} \Big[1 + e^{-2i\sigma^3 y} + (1-u) e^{-2i\sigma^3 y} \frac{\not\Pi}{m} + (1-u)\Big(\frac{1-u}{\Delta} + \frac{u}{\Delta} \frac{\sin y}{y} e^{-i\sigma^3 y} - e^{-2i\sigma^3 y} \Big) \frac{\not\Pi_\perp}{m} \Big] - (1+u) - (m + \not\Pi)\Big[\frac{1-u}{m} - 2imu(1-u^2)s \Big] \Big\}$$

$$y := eBsu$$

$$\phi := u(1-u)\left[m^2 - (\gamma\Pi)^2\right] + \frac{u}{y}\left[\beta - (1-u)y\right]\Pi_\perp^2 - u^2 \frac{e}{2}\sigma_{\mu\nu}F^{\mu\nu}$$

$$\Delta := (1-u)^2 + 2u(1-u)\sin y \cos y / y + u^2(\sin y / y)^2$$

(3.45) Spectral representation of the mass operator for B = 0

$$\Sigma_0(p) = -(\not p + m)^2 \frac{\alpha}{4\pi} \int_{\to m}^\infty \frac{dM}{M}\Big(1 - \frac{m^2}{M^2}\Big) \left\{ \frac{1 - \frac{2mM}{(M-m)^2}}{\not p + M - i\varepsilon} + \frac{1 + \frac{2mM}{(M+m)^2}}{\not p - M + i\varepsilon} \right\}$$

4th Section

(4.1') Polarization tensor in momentum representation

$$\Pi^{(2)}_{\mu\nu}(k) = -ie^2\, tr \int \frac{d^4p}{(2\pi)^4} \left[\gamma_\mu\, g(p)\, \gamma_\nu\, g(p-k)\right]$$

Definition of the polarization function

$$\Pi_{\mu\nu}(k) = (g_{\mu\nu}k^2 - k_\mu k_\nu)\Pi(k^2)$$

(4.32) Tsai's representation of the polarization tensor in a constant external magnetic field

$$\Pi_{\mu\nu}(k) = \frac{\alpha}{2\pi} \int_0^\infty \frac{ds}{s} \int_{-1}^{1} \frac{dv}{z} \{ e^{-is\varphi_0} [(g_{\mu\nu}k^2 - k_\mu k_\nu) N_0 -$$

$$- (g^{\parallel}_{\mu\nu} k_\parallel^2 - k_{\parallel\mu} k_{\parallel\nu}) N_1 + (g^{\perp}_{\mu\nu} k_\perp^2 - k_{\perp\mu} k_{\perp\nu}) N_2]$$

$$- e^{-ism^2} (1-v^2)(k^2 g_{\mu\nu} - k_\mu k_\nu)\}$$

$$\varphi_0 := m^2 + \frac{1}{4}(1-v^2) k_\parallel^2 + \frac{\cos zv - \cos z}{2z \sin z} k_\perp^2$$

$$N_0 := \frac{z}{\sin z}(\cos zv - v \cot z \sin zv)$$

$$N_1 := -z \cot z \left(1 - v^2 + \frac{v \sin zv}{\sin z}\right) + z \frac{\cos zv}{\sin z}$$

$$N_2 := -\frac{z \cos zv}{\sin z} + \frac{zv \cot z \sin zv}{\sin z} + \frac{2z(\cos zv - \cos z)}{\sin^3 z}$$

(4.34) Spectral representation of the polarization function for B = 0

$$\Pi(K^2) = -\frac{\alpha}{3\pi} K^2 \int_{4m^2}^{\infty} \frac{dM^2}{M^2}\left(1-\frac{4m^2}{M^2}\right)^{\frac{1}{2}}\left(1+\frac{2m^2}{M^2}\right)\frac{1}{K^2+M^2-i\varepsilon}$$

5th Section

(5.1) One-loop vacuum amplitude

$$\langle 0_+|0_-\rangle = \exp\{iW^{(1)}[A]\} = \exp\{i\int d^4x\, \mathcal{L}^{(1)}(x)\}$$

$$iW^{(1)}[A] = -\,\mathrm{Tr}\,\ln(1-e\not{A}G_+)^{-1}$$

$$= -\,\mathrm{Tr}\,\ln(G_+[A]/G_+[0])$$

(5.13)

$$i\,\frac{\partial\mathcal{L}^{(1)}(x)}{\partial m} = \mathrm{tr}\,G_+(x,x|A)$$

(5.23) Integral representation of the one-loop effective Lagrangian for a constant magnetic and vanishing electric field

$$\mathcal{L}_R^{(1)}(B) = \frac{1}{8\pi^2}\int_0^{\infty}\frac{ds}{s^3}e^{-im^2s}\left[(eBs)\cot(eBs)+\frac{1}{3}(eBs)^2-1\right]$$

(5.25)

$$\mathcal{L}_R^{(1)}(B) = \frac{-1}{8\pi^2}\int_0^{\infty}\frac{ds}{s^3}e^{-m^2s}\left[(eBs)\coth(eBs)-\frac{1}{3}(eBs)^2-1\right]$$

(5.26) Integral representation of the one-loop effective Lagrangian for a constant electric and vanishing magnetic field

$$\mathcal{L}_R^{(1)}(E) = \frac{-1}{8\pi^2}\int_0^\infty \frac{ds}{s^3} e^{-m^2 s}\left[(eEs)\cot(eEs) + \frac{1}{3}(eEs)^2 - 1\right]$$

(5.27) Representation of $L_R^{(1)}(B)$ by the Riemann Zeta-function

$$\mathcal{L}_R^{(1)}(B) = -\frac{1}{32\pi^2}\Big\{\left(2m^4 - 4m^2(eB) + \tfrac{4}{3}(eB)^2\right)\left[1 + \ln\frac{m^2}{2eB}\right]$$

$$+ 4m^2(eB) - 3m^4 - (4eB)^2\,\zeta'\!\left(-1, \frac{m^2}{2eB}\right)\Big\}$$

(5.29) asymptotic form

$$\mathcal{L}_R^{(1)}\left(\frac{eB}{m^2} \to \infty\right) = \frac{\alpha B^2}{6\pi}\left[\ln\frac{eB}{m^2} + 12\zeta'(-1) - 1 + \ln 2\right]$$

6th Section

(6.3) Definition of the zeta-function

$$\zeta_A(s) = \sum_n a_n^{-s}$$

(6.4) Definition of the determinant

$$\det A = \exp\left[-\zeta_A'(0)\right]$$

7th Section

(7.2) Vacuum amplitude

$$\langle 0_+ | 0_- \rangle_A = \exp\left\{-\frac{i}{2}\frac{\delta}{\delta J} D_+ \frac{\delta}{\delta J}\right\} \exp\left\{iW[A+J]\right\}\Big|_{J=0}$$

$$iW[A] \equiv -\operatorname{Tr}\ln(1 - e\not{A}\,G_+)^{-1}$$

(D.5) Functional derivatives of W[A]

$$i \frac{\delta W[A]}{\delta A^{\mu}(x)} = -e \, tr[\gamma_\mu G_+(x,x|A)]$$

$$i \frac{\delta^2 W[A]}{\delta A^{\mu_1}(x_1)\, \delta A^{\mu_2}(x_2)} = -e^2 \, tr[\gamma_{\mu_1} G_+(x_1,x_2|A) \gamma_{\mu_2} G_+(x_2,x_1|A)]$$

(7.9) Two-loop effective Lagrangian for constant fields

$$\mathcal{L}^{(2)} = \frac{e^2}{2} \int d^4x' \, tr[\gamma_\mu G_+(x,x'|A) \gamma_\nu G_+(x',x|A)] \, D_+^{\mu\nu}(x-x')$$

(7.10)

$$\mathcal{L}^{(2)} = \frac{e^2}{2} \int \frac{d^4p}{(2\pi)^4} \int \frac{d^4k}{(2\pi)^4} \, tr[\gamma_\mu g(p) \gamma^\mu g(p-k)] \, D_+(k)$$

(7.37) Integral representation for $L_R^{(2)}$

$$\mathcal{L}_R^{(2)}(B) = \frac{\alpha}{(4\pi)^3} \int_0^\infty \frac{ds}{s^3} \int_0^1 dv \, e^{-im^2 s} \left[K(z,v) - K_{02}(z,v) - \frac{f(z)}{1-v^2}\right]$$

$$- \frac{5}{6} \frac{3\alpha m}{4\pi} \frac{(-im)}{4\pi^2} \int_0^\infty \frac{ds}{s^2} e^{-im^2 s} \left[(eBs)\cot(eBs) + \tfrac{1}{3}(eBs)^2 - 1\right]$$

$$+ \frac{3\alpha}{16\pi^3} \int_0^\infty \frac{ds}{s^3} e^{-im^2 s} \ln(i\gamma m^2 s) \left[\frac{(eBs)^2}{\sin^2(eBs)} + (eBs)\cot(eBs) - 2\right]$$

$$z := eBs$$

$$f(z) := 24 \left[\frac{z^2}{\sin^2 z} + z\cot z - 2\right]$$

$$K_{02}(z,v) := \frac{48}{1-v^2} + 2z^2$$

$$K(z,v) := K_1(z,v) + K_2(z,v) \ln \frac{b}{a}(z,v)$$

$$K_1(z,v) := \frac{16z}{\sin z}\left[\frac{z \sin z \, (\cos zv - v\cot z \cdot \sin zv)}{(1-v^2)(\cos zv - \cos z)} + \frac{z(\cos z \cdot \sin z - z)}{2(\cos zv - \cos z) - (1-v^2) z \sin z}\right]$$

$$K_2(z,v) := \frac{16z^3}{\sin z} \frac{[2(\cos zv - \cos z) - (1-v^2)\cos z \cdot \sin^2 z]}{[2(\cos zv - \cos z) - (1-v^2)\, z \sin z]^2}$$

$$\frac{b}{a} := \frac{2(\cos zv - \cos z)}{(1-v^2)\, z \sin z}$$

(7.43) asymptotic form

$$\mathcal{L}_R^{(2)}\left(\frac{eB}{m^2} \to \infty\right) = \frac{\alpha^2 B^2}{8\pi^2} \ln \frac{eB}{m^2}$$

8th Section

(8.11) 't Hooft-Weinberg equation for $L_R = -\frac{1}{2} B^2 L$

$$[t\frac{\partial}{\partial t} - \beta_{tW}(\alpha_R)\alpha_R \frac{\partial}{\partial \alpha_R} + [1+\gamma_m(\alpha_R)]\varrho\frac{\partial}{\partial \varrho} + \beta_{tW}(\alpha_R)]\, L(t,\alpha_R,\varrho) = 0$$

$$t \equiv (eB)^{1/2}/\mu$$

$$\varrho \equiv m_R/\mu$$

$$\beta_{tW} \equiv \frac{\mu}{\alpha_R}\frac{\partial \alpha_R}{\partial \mu} = \frac{\mu}{Z_3}\frac{\partial Z_3}{\partial \mu}$$

(8.14) Callan-Symanzik equation for $\ell_{R\infty} = \frac{\mathcal{L}_R}{\mathcal{L}_R^{(0)}}\left(\frac{eB}{m^2} \to \infty\right)$:

$$[m^2\frac{\partial}{\partial m^2} + \frac{1}{2}\beta_{CS}(\alpha)(\alpha\frac{\partial}{\partial \alpha} - 1)]\, \ell_{R\infty}\left(\frac{eB}{m^2}, \alpha\right) = 0$$

$$\beta_{CS}(\alpha) \equiv \frac{m}{Z_3}\frac{\partial Z_3}{\partial m}$$

$$\beta_{CS}(\alpha) = \frac{2}{3}\left(\frac{\alpha}{\pi}\right) + \frac{1}{2}\left(\frac{\alpha}{\pi}\right)^2 - \frac{121}{144}\left(\frac{\alpha}{\pi}\right)^3 + O(\alpha^4)$$

References

[1] G. Mie, Ann.Phys. 37, 511 (1912); 39, 1 (1913), M. Born, L. Infeld, Proc.Roy.Soc. A 144, 425 (1934), W. Heisenberg, H. Euler, Z. Phys. 98, 714 (1936), V. Weisskopf, K. Danske Vidensk.Selsk.Mat.-fys.Medd. 14, No. 6 (1936), M. Greenman and F. Rohrlich, Phys.Rev. D8, 1103 (1973)

[2] S.L. Adler, Ann.Phys. (N.Y.) 67, 599 (1971)

[3] J. Schwinger, Phys.Rev. 82, 664 (1951)

[4] V.I. Ritus, JETP 42, 774 (1976)

[5] W. Dittrich, J.Phys. A, Vol. 10, No. 5, 833 (1977)

[6] W.-Y. Tsai, Phys.Rev. D10, 2699 (1974)

[7] W.-Y. Tsai, Phys.Rev. D10, 1342 (1974)

[8] W. Dittrich, J.Phys. A, Vol. 9, No. 7, 1171 (1976)

[9] K.-H. Zimmermann, Diplomarbeit, Univ. Tübingen (1978)

[10] W. Dittrich, W.-Y. Tsai, K.-H. Zimmermann, Phys.Rev. D19, 2929 (1979)

[11] S.R. Valluri, D. Lamm, W.J. Mielniczuk, Phys.Rev. D25, 2729 (1982)

[12] P. Ramond, "Field Theory", Benjamin Cummings, Reading/ Mass., (1981)

[13] G. Leibbrandt, Rev.Mod.Phys. 47, 849 (1975)

[14] J.C. Collins, Phys.Rev. D10, 1213 (1973)

[15] S.W. Hawking, Com.Math.Phys. 55, 133 (1977)

[16] G. Ghika, H. Visinescu, Nuovo Cimento 46A, 25 (1978)

[17] G.W. Gibbons, Phys.Letters 60A, 385 (1977)

[18] S. Weinberg, Phys.Rev. D9, 3320 (1974)

[19] W. Dittrich, Phys.Rev. D19, 2385 (1978)

[20] H.B.G. Casimir, Proc.kon.Ned.Akad.Wetenschap. 51, 793 (1948)

[21] E.G. Harris, "A Pedestrian Approach to Quantum Field Theory", J. Wiley, New York

[22] J. Schwinger, Letters in Math.Phys. 1, 43 (1975)

[23] B.V. Deryagin, I.I. Abrikosava, JETP 2, 73 (1956)

[24] R.P. Feynman, A.R. Hibbs, "Quantum Mechanics and Path Integrals", McGraw-Hill, New York

[25] E.S. Abers, B.W. Lee, Phys.Reports 9C, 1 (1973

[26] J.C. Taylor, "Gauge Theories of Weak Interactions", Cambridge University Press (1976)

[27] R. Jackiw, Phys.Rev. D9, 1686 (1974)

[28] S. Coleman, E. Weinberg, Phys.Rev. D7, 1888 (1973)

[29] J. Schwinger, "Particles, Sources and Fields", Addison-Wesley Co., Reading/Mass. (1970/1973)

[30] W. Dittrich, Fortschritte der Physik 26, 289 (1978)

[31] J.Z. Kaminski, Act.Phys.Austr. 53, 287 (1981)

[32] R. Coquereaux, Ann.Phys. 125 (1980)

[33] C. Nash, "Relativistic Quantum Fields", Academic Press (1978)

[34] F. Ravndal, "Scaling and Renormalization Groups", Nordita (1976)

[35] J.C. Collins, A.J. Macfarlane, Phys.Rev. D10, 1201 (1974

[36] I.S. Gradstein, I.M. Ryshik, "Tables of Series, Products and Integrals", Harri Deutsch, Thun, Frankfurt/M. (1981)

[37] J.D. Bjorken, S.D. Drell, "Relativistische Quantenmechanik", "Relativistische Quantenfeldtheorie", Bibliograph. Inst., Mannheim

[38] V. Schanbacher, Diplomarbeit, Univ. Tübingen (1979)

[39] K. Nishijima, "Fields and Particles", W.A. Benjamin (1969)

[40] W.J. Mielniczuk, J.Phys. A15, 2905 (1982)

[41] H.M. Fried, "Functional Methods and Models in Quantum Field Theory", MIT Press, Cambridge (1972)

[42] E. de Rafael, J.L. Rosner, Ann.Phys. 82 369 (1974)

[43] M.R. Brown, M.J. Duff, Phys.Rev. D11, 2124 (1975)

[44] A. Messiah, "Quantenmechanik", de Gruyter (1976)

[45] M.J. Teper, "Instantons θ-Vacua, Confinement,...", Rutherford Laboratory (1980)

[46] R.J. Hughes, Nucl.Phys. B186, 376 (1981)

[47] W. Dittrich, "Lectures on Schwinger's Source Methods in Quantum Field Theory", Univ. Tübingen, WS 1975/76

[48] N.D. Birrell, P.C.W. Davies, "Quantum Fields in Curved Space", Cambridge Univ. Press (1982)

[49] S. Weinberg, "Gravitation and Cosmology", John Wiley (1972)

[50] S.L. Adler, Rev.Mod.Phys. 54, 729 (1982)

[51] P. Becher, M. Böhm, H. Joos, "Eichtheorien der starken und elektroschwachen Wechselwirkung"; Teubner, Stuttgart (1981)

[52] H. Mitter, Physik.Blätter 24, 159 (1968); 24, 348 (1968)

[53] C. Itzykson, J.B. Zuber, "Quantum Field Theory", McGraw-Hill, New York (1980)

[54] J.S. Dowker, R. Critchley, Phys.Rev. D13, 3224 (1976)

[55] Y. Nambu, Phys.Rev.Lett. 4, 380 (1960), J. Goldstone, Nuovo Cimento 19, 154 (1961)

[56] R. Jackiw, Lectures given at Les Houches, 1983, MIT print CTP 1108

[57] H. Pagels, E. Tomboulis, Nucl.Phys. B143, 485 (1978)

[58] A.B. Migdal, Nucl.Phys. B52, 483 (1973)

[59] S.L. Adler, "Dynamical Applications of the gauge-invariant effective action formalism", to appear in the Bryce DeWitt 60th Birthday Quantum Gravity Volume

[60] F.J. Yndurâin, "Quantum Chromodynamics", Springer (1983)

[61] W. Dittrich, M. Reuter, Phys.Lett. 128B, 321 (1983)

[62] S.L. Adler, "Non-Abelian Statics", Talk presented at the Maxwell Symposium, Amherst (1981)

[63] S.L. Adler, Phys.Rev. D23, 2905 (1981); Phys.Lett. 110B, 302 (1982); ibid. 113B, 405 (1982); ibid. 117B, 91 (1982); Nucl.Phys. B217, 381 (1983), S.L. Adler in workshop "Non-Perturbative QCD", Stillwater, Oklahoma 1983; K.A. Milton and M.A. Samuel, Eds. Progress in Physics Vol. 8, Birkhäuser Boston-Basel-Stuttgart, 1983

[64] H.B. Nielsen in Proceedings of the 3rd Adriatic Summer Meeting on Particle Physics, I. Andrić et al., eds. North-Holland Publishing Co., 1981

M. D. Scadron

Advanced Quantum Theory

and Its Applications Through Feynman Diagrams

Corrected 2nd printing. 1981. 78 figures.
XIV, 386 pages. (Texts and Monographs in Physics)
ISBN 3-540-10970-6

Contents: Transformation Theory: Introduction. Transformations in Space. Transformations in Space-Time. Boson Wave Equations. Spin-½ Dirac Equation. Discrete Symmetries. - Scattering Theory: Formal Theory of Scattering. Simple Scattering Dynamics. Nonrelativistic Perturbation Theory. - Covariant Feynman Diagrams: Covariant Feynman Rules. Lowest-Order Electromagnetic Interactions. Low-Energy Strong Interactions. Lowest-Order Weak Interactions. Lowest-Order Gravitational Interactions. Higher-Order Covariant Feynman Diagrams. - Problems. - Appendices. - Bibliography. - Index.

F. J. Ynduráin

Quantum Chromodynamics

An Introduction to the Theory of Quarks and Gluons

1983. XI, 227 pages. (Texts and Monographs in Physics). ISBN 3-540-11752-0

Contents: Generalities. - QCD as a Field Theory. - Deep Inelastic Processes. - Quark Masses, PCAC, Chiral Dynamics, and the QCD Vacuum. - Functional Methods, Nonperturbative Solution. - References. - Index.

Springer-Verlag
Berlin
Heidelberg
New York
Tokyo

Lecture Notes in Physics

For information about Vols. 1–151, please contact your bookseller or Springer-Verlag.

Vol. 152: Physics of Narrow Gap Semiconductors. Proceedings, 1981. Edited by E. Gornik, H. Heinrich and L. Palmetshofer. XIII, 485 pages. 1982.

Vol. 153: Mathematical Problems in Theoretical Physics. Proceedings, 1981. Edited by R. Schrader, R. Seiler, and D.A. Uhlenbrock. XII, 429 pages. 1982.

Vol. 154: Macroscopic Properties of Disordered Media. Proceedings, 1981. Edited by R. Burridge, S. Childress, and G. Papanicolaou. VII, 307 pages. 1982.

Vol. 155: Quantum Optics. Proceedings, 1981. Edited by C.A. Engelbrecht. VIII, 329 pages. 1982.

Vol. 156: Resonances in Heavy Ion Reactions. Proceedings, 1981. Edited by K.A. Eberhard. XII, 448 pages. 1982.

Vol. 157: P. Niyogi, Integral Equation Method in Transonic Flow. XI, 189 pages. 1982.

Vol. 158: Dynamics of Nuclear Fission and Related Collective Phenomena. Proceedings, 1981. Edited by P. David, T. Mayer-Kuckuk, and A. van der Woude. X, 462 pages. 1982.

Vol. 159: E. Seiler, Gauge Theories as a Problem of Constructive Quantum Field Theory and Statistical Mechanics. V, 192 pages. 1982.

Vol. 160: Unified Theories of Elementary Particles. Critical Assessment and Prospects. Proceedings, 1981. Edited by P. Breitenlohner and H.P. Dürr. VI, 217 pages. 1982.

Vol. 161: Interacting Bosons in Nuclei. Proceedings, 1981. Edited by J.S. Dehesa, J.M.G. Gomez, and J. Ros. V, 209 pages. 1982.

Vol. 162: Relativistic Action at a Distance: Classical and Quantum Aspects. Proceedings, 1981. Edited by J. Llosa. X, 263 pages. 1982.

Vol. 163: J. S. Darrozes, C. Francois, Mécanique des Fluides Incompressibles. XIX, 459 pages. 1982.

Vol. 164: Stability of Thermodynamic Systems. Proceedings, 1981. Edited by J. Casas-Vázquez and G. Lebon. VII, 321 pages. 1982.

Vol. 165: N. Mukunda, H. van Dam, L.C. Biedenharn, Relativistic Models of Extended Hadrons Obeying a Mass-Spin Trajectory Constraint. Edited by A. Böhm and J.D. Dollard. VI, 163 pages. 1982.

Vol. 166: Computer Simulation of Solids. Edited by C.R.A. Catlow and W.C. Mackrodt. XII, 320 pages. 1982.

Vol. 167: G. Fieck, Symmetry of Polycentric Systems. VI, 137 pages, 1982.

Vol. 168: Heavy-Ion Collisions. Proceedings, 1982. Edited by G. Madurga and M. Lozano. VI, 429 pages. 1982.

Vol. 169: K. Sundermeyer, Constrained Dynamics. IV, 318 pages. 1982.

Vol. 170: Eighth International Conference on Numerical Methods in Fluid Dynamics. Proceedings, 1982. Edited by E. Krause. X, 569 pages. 1982.

Vol. 171: Time-Dependent Hartree-Fock and Beyond. Proceedings, 1982. Edited by K. Goeke and P.-G. Reinhard. VIII, 426 pages. 1982.

Vol. 172: Ionic Liquids, Molten Salts and Polyelectrolytes. Proceedings, 1982. Edited by K.-H. Bennemann, F. Brouers, and D. Quitmann. VII, 253 pages. 1982.

Vol. 173: Stochastic Processes in Quantum Theory and Statistical Physics. Proceedings, 1981. Edited by S. Albeverio, Ph. Combe, and M. Sirugue-Collin. VIII, 337 pages. 1982.

Vol. 174: A. Kadić, D.G.B. Edelen, A Gauge Theory of Dislocations and Disclinations. VII, 290 pages. 1983.

Vol. 175: Defect Complexes in Semiconductor Structures. Proceedings, 1982. Edited by J. Giber, F. Beleznay, J. C. Szép, and J. László. VI, 308 pages. 1983.

Vol. 176: Gauge Theory and Gravitation. Proceedings, 1982. Edited by K. Kikkawa, N. Nakanishi, and H. Nariai. X, 316 pages. 1983.

Vol. 177: Application of High Magnetic Fields in Semiconductor Physics. Proceedings, 1982. Edited by G. Landwehr. XII, 552 pages. 1983.

Vol. 178: Detectors in Heavy-Ion Reactions. Proceedings, 1982. Edited by W. von Oertzen. VIII, 258 pages. 1983.

Vol. 179: Dynamical Systems and Chaos. Proceedings, 1982. Edited by L. Garrido. XIV, 298 pages. 1983.

Vol. 180: Group Theoretical Methods in Physics. Proceedings, 1982. Edited by M. Serdaroğlu and E. İnönü. XI, 569 pages. 1983.

Vol. 181: Gauge Theories of the Eighties. Proceedings, 1982. Edited by R. Raitio and J. Lindfors. V, 644 pages. 1983.

Vol. 182: Laser Physics. Proceedings, 1983. Edited by J. D. Harvey and D. F. Walls. V, 263 pages. 1983.

Vol. 183: J. D. Gunton, M. Droz, Introduction to the Theory of Metastable and Unstable States. VI, 140 pages. 1983.

Vol. 184: Stochastic Processes – Formalism and Applications. Proceedings, 1982. Edited by G. S. Agarwal and S. Dattagupta. VI, 324 pages. 1983.

Vol. 185: H. N. Shirer, R. Wells, Mathematical Structure of the Singularities at the Transitions between Steady States in Hydrodynamic Systems. XI, 276 pages. 1983.

Vol. 186: Critical Phenomena. Proceedings, 1982. Edited by F.J.W. Hahne. VII, 353 pages. 1983.

Vol. 187: Density Functional Theory. Edited by J. Keller and J.L. Gázquez. V, 301 pages. 1983.

Vol. 188: A. P. Balachandran, G. Marmo, B.-S. Skagerstam, A. Stern, Gauge Symmetries and Fibre Bundles. IV, 140 pages. 1983.

Vol. 189: Nonlinear Phenomena. Proceedings, 1982. Edited by K. B. Wolf. XII, 453 pages. 1983.

Vol. 190: K. Kraus, States, Effects, and Operations. Edited by A. Böhm, J. W. Dollard and W. H. Wootters. IX, 151 pages. 1983.

Vol. 191: Photon Photon Collisions. Proceedings, 1983. Edited by Ch. Berger. V, 417 pages. 1983.

Vol. 192: Heidelberg Colloquium on Spin Glasses. Proceedings, 1983. Edited by J. L. van Hemmen and I. Morgenstern. VII, 356 pages. 1983.

Vol. 193: Cool Stars, Stellar Systems, and the Sun. Proceedings, 1983. Edited by S. L. Balliunas and L. Hartmann. VII, 364 pages. 1984.

Vol. 194: P. Pascual, R. Tarrach, QCD: Renormalization for the Practitioner. V, 277 pages. 1984.

Lecture Notes in Physics

Vol. 195: Trends and Applications of Pure Mathematics to Mechanics. Proceedings, 1983. Edited by P. G. Ciarlet and M. Roseau. V, 422 pages. 1984.

Vol. 196: WOPPLOT 83. Parallel Processing: Logic, Organization and Technology. Proceedings, 1983. Edited by J. Becker and I. Eisele. V, 189 pages. 1984.

Vol. 197: Quarks and Nuclear Structure. Proceedings, 1983. Edited by K. Bleuler. VIII, 414 pages. 1984.

Vol. 198: Recent Progress in Many-Body Theories. Proceedings, 1983. Edited by H. Kümmel and M. L. Ristig. IX, 422 pages. 1984.

Vol. 199: Recent Developments in Nonequilibrium Thermodynamics. Proceedings, 1983. Edited by J. Casas-Vázquez, D. Jou and G. Lebon. XIII, 485 pages. 1984.

Vol. 200: H.D. Zeh, Die Physik der Zeitrichtung. V, 86 Seiten. 1984.

Vol. 201: Group Theoretical Methods in Physics. Proceedings, 1983. Edited by G. Denardo, G. Ghirardi and T. Weber. XXXVII, 518 pages. 1984.

Vol. 202: Asymptotic Behavior of Mass and Spacetime Geometry. Proceedings, 1983. Edited by F. J. Flaherty. VI, 213 pages. 1984.

Vol. 203: C. Marchioro, M. Pulvirenti, Vortex Methods in Two-Dimensional Fluid Dynamics. III, 137 pages. 1984.

Vol. 204: Y. Waseda, Novel Application of Anomalous (Resonance) X-Ray Scattering for Structural Characterization of Disordered Materials. VI, 183 pages. 1984.

Vol. 205: Solutions of Einstein's Equations: Techniques and Results. Proceedings, 1983. Edited by C. Hoenselaers and W. Dietz. VI, 439 pages. 1984.

Vol. 206: Static Critical Phenomena in Inhomogeneous Systems. Edited by A. Pękalski and J. Sznajd. Proceedings, 1984. VIII, 358 pages. 1984.

Vol. 207: S. W. Koch, Dynamics of First-Order Phase Transitions in Equilibrium and Nonequilibrium Systems. III, 148 pages. 1984.

Vol. 208: Supersymmetry and Supergravity/Nonperturbative QCD. Proceedings, 1984. Edited by P. Roy and V. Singh. V, 389 pages. 1984.

Vol. 209: Mathematical and Computational Methods in Nuclear Physics. Proceedings, 1983. Edited by J. S. Dehesa, J. M. G. Gomez and A. Polls. V, 276 pages. 1984.

Vol. 210: Cellular Structures in Instabilities. Proceedings, 1983. Edited by J. E. Wesfreid and S. Zaleski. VI, 389 pages. 1984.

Vol. 211: Resonances – Models and Phenomena. Proceedings, 1984. Edited by S. Albeverio, L. S. Ferreira and L. Streit. VI, 369 pages. 1984.

Vol. 212: Gravitation, Geometry and Relativistic Physics. Proceedings, 1984. Edited by Laboratoire "Gravitation et Cosmologie Relativistes", Université Pierre et Marie Curie et C.N.R.S., Institut Henri Poincaré, Paris. VI, 336 pages. 1984.

Vol. 213: Forward Electron Ejection in Ion Collisions. Proceedings, 1984. Edited by K.O. Groeneveld, W. Meckbach and I.A. Sellin. VII, 165 pages. 1984.

Vol. 214: H. Moraal, Classical, Discrete Spin Models. VII, 251 pages. 1984.

Vol. 215: Computing in Accelerator Design and Operation. Proceedings, 1983. Edited by W. Busse and R. Zelazny. XII, 574 pages. 1984.

Vol. 216: Applications of Field Theory to Statistical Mechanics. Proceedings, 1984. Edited by L. Garrido. VIII, 352 pages. 1985.

Vol. 217: Charge Density Waves in Solids. Proceedings, 1984. Edited by Gy. Hutiray and J. Sólyom. XIV, 541 pages. 1985.

Vol. 218: Ninth International Conference on Numerical Methods in Fluid Dynamics. Edited by Soubbaramayer and J. P. Boujot. X, 612 pages. 1985.

Vol. 219: Fusion Reactions Below the Coulomb Barrier. Proceedings, 1984. Edited by S.G. Steadman. VII, 351 pages. 1985.

Vol. 220: W. Dittrich, M. Reuter, Effective Lagrangians in Quantum Electrodynamics. V, 244 pages. 1985.

Selected Issues from

Lecture Notes in Mathematics

Vol. 932: Analytic Theory of Continued Fractions. Proceedings, 1981. Edited by W.B. Jones, W.J. Thron, and H. Waadeland. VI, 240 pages. 1982.

Vol. 934: M. Sakai, Quadrature Domains. IV, 133 pages. 1982.

Vol. 935: R. Sot, Simple Morphisms in Algebraic Geometry. IV, 146 pages. 1982.

Vol. 936: S.M. Khaleelulla, Counterexamples in Topological Vector Spaces. XXI, 179 pages. 1982.

Vol. 937: E. Combet, Intégrales Exponentielles. VIII, 114 pages. 1982.

Vol. 938: Number Theory. Proceedings, 1981. Edited by K. Alladi. IX, 177 pages. 1982.

Vol. 942: Theory and Applications of Singular Perturbations. Proceedings, 1981. Edited by W. Eckhaus and E.M. de Jager. V, 363 pages. 1982.

Vol. 953: Iterative Solution of Nonlinear Systems of Equations. Proceedings, 1982. Edited by R. Ansorge, Th. Meis, and W. Törnig. VII, 202 pages. 1982.

Vol. 956: Group Actions and Vector Fields. Proceedings, 1981. Edited by J.B. Carrell. V, 144 pages. 1982.

Vol. 957: Differential Equations. Proceedings, 1981. Edited by D.G. de Figueiredo. VIII, 301 pages. 1982.

Vol. 963: R. Nottrot, Optimal Processes on Manifolds. VI, 124 pages. 1982.

Vol. 964: Ordinary and Partial Differential Equations. Proceedings, 1982. Edited by W.N. Everitt and B.D. Sleeman. XVIII, 726 pages. 1982.

Vol. 968: Numerical Integration of Differential Equations and Large Linear Systems. Proceedings, 1980. Edited by J. Hinze. VI, 412 pages. 1982.

Vol. 970: Twistor Geometry and Non-Linear Systems. Proceedings, 1980. Edited by H.-D. Doebner and T.D. Palev. V, 216 pages. 1982.

Vol. 972: Nonlinear Filtering and Stochastic Control. Proceedings, 1981. Edited by S.K. Mitter and A. Moro. VIII, 297 pages. 1983.

Vol. 978: J. Ławrynowicz, J. Krzyż, Quasiconformal Mappings in the Plane. VI, 177 pages. 1983.

Vol. 979: Mathematical Theories of Optimization. Proceedings, 1981. Edited by J.P. Cecconi and T. Zolezzi. V, 268 pages. 1983.

Vol. 982: Stability Problems for Stochastic Models. Proceedings, 1982. Edited by V.V. Kalashnikov and V.M. Zolotarev. XVII, 295 pages. 1983.

Vol. 989: A.B. Mingarelli, Volterra-Stieltjes Integral Equations and Generalized Ordinary Differential Expressions. XIV, 318 pages. 1983.

Vol. 994: J.-L. Journé, Calderón-Zygmund Operators, Pseudo-Differential Operators and the Cauchy Integral of Calderón. VI, 129 pages. 1983.

Vol. 999: C. Preston, Iterates of Maps on an Interval. VII, 205 pages. 1983.

Vol. 1000: H. Hopf, Differential Geometry in the Large, VII, 184 pages. 1983.

Vol. 1003: J. Schmets, Spaces of Vector-Valued Continuous Functions. VI, 117 pages. 1983.

Vol. 1005: Numerical Methods. Proceedings, 1982. Edited by V. Pereyra and A. Reinoza. V, 296 pages. 1983.

Vol. 1007: Geometric Dynamics. Proceedings, 1981. Edited by J. Palis Jr. IX, 827 pages. 1983.

Vol. 1015: Equations différentielles et systèmes de Pfaff dans le champ complexe – II. Seminar. Edited by R. Gérard et J.P. Ramis. V, 411 pages. 1983.

Vol. 1021: Probability Theory and Mathematical Statistics. Proceedings, 1982. Edited by K. Itô and J.V. Prokhorov. VIII, 747 pages. 1983.

Vol. 1031: Dynamics and Processes. Proceedings, 1981. Edited by Ph. Blanchard and L. Streit. IX, 213 pages. 1983.

Vol. 1032: Ordinary Differential Equations and Operators. Proceedings, 1982. Edited by W.N. Everitt and R.T. Lewis. XV, 521 pages. 1983.

Vol. 1035: The Mathematics and Physics of Disordered Media. Proceedings, 1983. Edited by B.D. Hughes and B.W. Ninham. VII, 432 pages. 1983.

Vol. 1037: Non-linear Partial Differential Operators and Quantization Procedures. Proceedings, 1981. Edited by S.I. Andersson and H.-D. Doebner. VII, 334 pages. 1983.

Vol. 1041: Lie Group Representations II. Proceedings 1982–1983. Edited by R. Herb, S. Kudla, R. Lipsman and J. Rosenberg. IX, 340 pages. 1984.

Vol. 1045: Differential Geometry. Proceedings, 1982. Edited by A.M. Naveira. VIII, 194 pages. 1984.

Vol. 1047: Fluid Dynamics. Seminar, 1982. Edited by H. Beirão da Veiga. VII, 193 pages. 1984.

Vol. 1048: Kinetic Theories and the Boltzmann Equation. Seminar, 1981. Edited by C. Cercignani. VII, 248 pages. 1984.

Vol. 1049: B. Iochum, Cônes autopolaires et algèbres de Jordan. VI, 247 pages. 1984.

Vol. 1054: V. Thomée, Galerkin Finite Element Methods for Parabolic Problems. VII, 237 pages. 1984.

Vol. 1055: Quantum Probability and Applications to the Quantum Theory of Irreversible Processes. Proceedings, 1982. Edited by L. Accardi, A. Frigerio and V. Gorini. VI, 411 pages. 1984.

Vol. 1057: Bifurcation Theory and Applications. Seminar, 1983. Edited by L. Salvadori. VII, 233 pages. 1984.

Vol. 1058: B. Aulbach, Continuous and Discrete Dynamics near Manifolds of Equilibria. IX, 142 pages. 1984.

Vol. 1059: Séminaire de Probabilités XVIII, 1982/83. Proceedings. Edité par J. Azéma et M. Yor. IV, 518 pages. 1984.

Vol. 1063: Orienting Polymers. Proceedings, 1983. Edited by J.L. Ericksen. VII, 166 pages. 1984.

Vol. 1065: A. Cuyt, Padé Approximants for Operators: Theory and Applications. IX, 138 pages. 1984.

Vol. 1066: Numerical Analysis. Proceedings, 1983. Edited by D.F. Griffiths. XI, 275 pages. 1984.

Vol. 1071: Padé Approximation and its Applications, Bad Honnef 1983. Prodeedings Edited by H. Werner and H.J. Bünger. VI, 264 pages. 1984.

Vol. 1072: F. Rothe, Global Solutions of Reaction-Diffusion Systems. V, 216 pages. 1984.

Vol. 1085: G.K. Immink, Asymptotics of Analytic Difference Equations. V, 134 pages. 1984.

Vol. 1086: Sensitivity of Functionals with Applications to Engineering Sciences. Proceedings, 1983. Edited by V. Komkov. V, 130 pages. 1984

Vol. 1100: V. Ivrii, The Precise Spectral Asymptotics for Elliptic Operators Acting in Fiberings over Manifolds with Boundary. V, 237 pages. 1984.